BREAKING ALL THE RULES

THE INSIDE STORY OF THE NEW SPACE RACE

JIM CANTRELL

ISBN 978-1-960546-94-4 (paperback)
ISBN 978-1-960546-95-1 (hardcover)
ISBN 978-1-960546-96-8 (digital)

Printed in the United States of America

*"The open road still softly calls us,
like a nearly forgotten song of childhood"*

Carl Sagan, The Pale Blue Dot

For Angela, who toiled alongside me for eight long years while I completed this work, woke me from nightmares of being trapped in Russia, stood behind me in good times and bad, and loved me every step of the way.

CONTENTS

CHAPTER 1

THE OPEN ROAD

Humanity has always been on the move. It is how we survive and this instinct lives in our DNA. We are all space travelers. We whirl through the cold dark cosmos at some 10,000 MPH on our spaceship called Earth. Yet, those restless souls on this planet yearn to leave it to find new worlds. This is their story. This is the story of a Revolution that has yet to fully play out.

Humankind has been Earthbound for its entire existence. Most of humanity is content to remain that way. However, something deep within our DNA calls for us to wander and explore. Carl Sagan termed it as something akin to a "long lost song from childhood that gently calls us from our past." Carl was deeply in touch with this part of our human spirit. We are descended from hunters and gatherers. This is in our DNA; we must continually move and expand as a species to remain who we are. Indeed, we change over time as a society and become more settled. Still, there remains a minority of the population that is driven by this deep yearning that we can barely articulate. It's a yearning to see what's around the next corner, to explore new vistas, and to challenge the accepted norms of behavior and thought. These are the rogue humans, and they are a significant force of change, advancement, and occasional chaos..

Today we stand at the edge of something that occurs only once in a lifetime or perhaps even once in a thousand years: a new frontier. That frontier is space and its economic promise—even salvation for our human race. In much the same way that Spaniards stood on the shores of Spain, watching the Galleons set off to the last great frontier of the New World, today we witness the first steps of mankind off the Earth's surface into a much more immense and nearly infinite frontier. Part of our motivation to spread our human wings into this new frontier is economic, and part of it is much deeper, like the urge to find what is around the next hill or over the next ocean. As with other Frontiers, the story of those early explorers is fascinating and equally instructive to understand some primary motivations of the human species. The Space Race is no different and is arguably an even bigger story than the New World and its settlement. While we are still at the very early stages of this adventure, it's clear that numerous factions are forming around economic exploitation of space, national security concerns, and even dreams of utopia with libertarian settlements on places like Mars.

The original Space Race was a contest between nation-states, the United States of America, and the Soviet Union. It was a competition of giants where the fate of the world seemed to depend on the outcome. The Space Race embodied the Cold War of the 1960's and all that we found meaningful in our nations. It also served as a proxy of military, technical, economic, and moral superiority in a battle for the hearts and minds of the world. Capitalism versus Communism battling on the world stage was something that the world watched closely, and the Space Race was the most visible contest on display in the 1960s. The Space Race culminated in the first landing of a human on the lunar surface in 1969 by the U.S. This was not simply an American victory. This was a victory for all of mankind as these were our first baby steps off the Earth and into the cosmos, and the rest of the world sensed this too. Despite this massive human achievement, Harrison Schmidt, the last astronaut to walk on the Moon, stepped off the surface a mere four years later in 1972 and flew back to Earth to very little fanfare.

No human has yet to return to the Moon since Apollo. At the time, most people envisioned the Apollo landings as the beginning of a long trek to settle the solar system and, later, the galaxies in the distance. We all expected our government to continue pushing ever deeper into the cosmos. Yet over the following decades, we remained tethered on our beautiful terraqueous blue globe called Earth, seemingly bound here as we have been for millennia. Sure, we ventured from its surface to orbit aboard magnificent machines that became mere distractions from what was once the great Space Race. However, the promise of Apollo was broken, with no real prospect for anything new to happen in our lifetimes.

Meanwhile, discontent was growing and stirring among those for whom the promise of Apollo still burned brightly inside them. For those who remembered this dream and who believed that humanity could and should venture forth into the cosmos, they refused to ever give up on that idea. These people came from all walks of life. They were the always restless builders, the engineers, the visionaries, and later the billionaires who eventually came together in the most unlikely ways to reignite the promise of Apollo.

Harrison Schmidt was the last US Astronaut to step on the Moon in this mission from 1972. Image Credit NASA

Forty years later, a New Space Race is on; this time, it's no longer the sole domain of nation-states. It's a race between entrepreneurs, billionaires, and visionaries with ideas to change how humanity travels to space and uses it for everyday life. This is the new Space Race, and it embodies the spirit and promise of Apollo as only a new generation could. Most people barely noticed as the space industry transitioned from a government-dominated, highly regulated monopoly to a dynamic commercial sector. Today it is a $400B a year industry, and it's headed to becoming a two trillion-dollar industry over the next decade by most estimates. SpaceX broke the mold several times over as they developed the world's first private rockets capable of sending satellites to space and later humans - replacing the aged-out NASA shuttle. In 2017 alone, more than $1.5B of venture capital was invested in this new industry, which compares with $23B for NASA's 2021 budget, making it one of the hottest investment areas next to Artificial Intelligence and Self Driving Cars. All this is happening right under the noses of most financial analysts, industry titans, and government officials. The great part of this story is how it came to be and where it will go. It's a story that has much in common with that of Apollo yet touches on who we are as individuals, as Americans, and as citizens of the world. It's the story of all of us, from ordinary people to extraordinary billionaires. It's a story as compelling as it is fascinating and essential.

The modern Space Race can trace its roots to the end of our last massively destructive war: World War II. Efforts undertaken to arm the United States to fight as part of World War II are almost beyond criticism in American politics. It may be surprising to many of us then that the policies and efforts employed over 70 years ago still affect our industry today and are in many ways at the heart of the massive transformation of the aerospace industry away from defeating a massively armed geopolitical foe to creating economic value in space.

Following World War One, the US military was anxious to demobilize its forces rapidly, as it had done after every previous war. By 1920, many Americans sought a return to quieter times and more traditional values. Politicians were also weary and carried their constituents' sentiments to the House floor. The result was two decades of meager investment in military readiness and technology. During this period, the U.S. military relied upon advances in the commercial industry at large. It adopted advances in aviation and electronics to meet its mission requirements as little military-funded technology development was to be had.

On the eve of U.S. involvement in World War II, with war already raging in Europe, the U.S. military began rearming and supplying its allies in Europe to win against a technologically superior German army and air force. The priority for military funding in the early 1940s was building enough armaments to meet the challenge of Nazi Germany. As the war progressed, new military thinking emerged to develop technology in response to German war technologies and their effectiveness on the battlefield. While sheer numbers of tanks, soldiers, planes, and logistics eventually won the war in Europe, the development of the atomic bomb, radar, the jet engine, ballistic missiles, and supersonic aircraft are legacies of World War II technology developments that would shape the next 65 years of government spending.

Today, it's hard to imagine the sheer magnitude of the industrial efforts to manufacture armaments to meet our needs in the World War II theater. Automobile factories ceased manufacturing passenger cars to free up capacity to manufacture tanks, aircraft, and weaponry on behalf of the federal government. Every industrial power that the U.S. government could use for war material production was employed. The government commandeered U.S. industry to win the war in Europe and Asia. Given the Great Depression that preceded this era, nobody complained about having jobs and income to feed their families while the nation was at war. U.S. government debt rose to historical levels to fund this expanded production to levels comparable to what we see today.

After the war ended, the industrial capacity was converted back to civilian production. However, the scientific treasure recovered from Nazi Germany and the evolving geopolitical threat from the Soviet Union fueled military-funded technology development. This had the effect of leaving a portion of the industrial conversion permanently in place. Companies such as Hughes and General Electric maintained a large postwar research-and-development (R&D) base to develop new weapons systems to remain one step ahead of the Soviets. In this sense, the industrial policy of the World War II era U.S. military never really ended but evolved to fit the Cold War. NASA sprang into existence to contest the Soviets' dominance in space exploration.

This unprecedented level of funding continued unabated through three more decades. Its final crescendo was the Reagan military buildup of the 1980s, when missile defense was developed, and the nation rearmed in a manner reminiscent of World War II. In 1991, the Soviet Union finally collapsed, ending the raison d'etre for the World War II industrial model that was again employed during the Reagan era. Saddam Hussein and his Iraqi army became the eventual and unfortunate recipients of a generation of weapons designed to destroy the Soviet army, adding an exclamation point to this period of history.

For the two decades following the demise of the Soviet Union, government funding struggled to find justification without the clear and present danger of an enemy such as the Soviets. The nation found that enemy on Sept. 11, 2001, and a new round of spending ensued. However, the United States had already accumulated an enormous debt by this point in history. The two simultaneous Middle East wars ran up trillions of dollars in additional debt the nation could not afford. The prosecution of the Iraq and Afghan wars continued. New technology funding flowed unabated until 2008 when a burgeoning private debt crisis that eerily mirrored the government debt crisis exploded and plunged the nation into an economic crisis not seen since the Great Depression. Many of us in the industry sensed that something fundamental was about to change when this crisis broke out, but few appreciated the profound change it brought.

The economic crisis that plagued the nation for ten years, starting in 2008, finally broke the U.S. World War II industrial model of government-commercial collaboration. The U.S. government is experiencing unprecedented budget deficits and can ill afford to continue spending vast amounts of money it does not have on new technology development. The dreaded term "sequestration" and the budget austerity it implies are a force here to stay.

Natural sociological and economic forces are forcing NASA and the defense technology complex to return to the model they had before World War II, in which government-funded R&D was indeed sparse. Before World War II, the U.S. government was forced to rely upon inventors and technology developed in the private sector and adopted to the military and government needs.

Without casting judgment on the policies of the past, World War II and the extended Cold War that followed turned the natural economic order and U.S. industrial model upside-down, where entire industries were converted into arms of the U.S. military and the U.S. government itself. While this was necessary to win the war, the reverse transformation, or demobilization, was never achieved. During the Cold War, our post-World War II global economic dominance was partly due to the fact that we had destroyed most of the world's industrial capacity, and we could dominate industrial-economic spheres for the next 50 years. The Cold War was underwritten by this economic dominance and allowed the World War II industrial model to remain intact. The financial collapse of 2008 was inevitable when overspending in the United States and the world's reindustrialization caught up. The permanent loss in tax revenue from the market housing bubble collapse, in addition to recent massive COVID excused spending bills, puts pressure on spending, and debt servicing is multiplying this pressure. The U.S. government has no way out in the long run other than cutting spending.

So, what should we expect in the next ten years? I am optimistic and believe in the spirit and power of capitalism and its ability to deploy capital, innovate, and produce value efficiently. Our aerospace industry will change and adapt to this new reality, and the U.S. government will find new ways to harness the more efficient capital deployment of the private sector.

Witness a superb example: SpaceX. By 2010, SpaceX had spent less than $1 billion in capital since its founding in 2001, and had launched five successful, Evolved Expendable Launch Vehicle-class vehicles (Falcon 9), five Falcon 1 vehicles, and four Dragon crew capsules, and built three launch pads. A part of this capital came from the U.S. government (about $600 million). Still, it was implemented by the private sector, and its deployment was undoubtedly a pure capitalist approach. Today, SpaceX is routinely launching astronauts, government and private, into space, all on this original platform.

Without entering all the arguments about crew safety and standards, it's hard to argue that this is not a more efficient deployment of capital than that of the Constellation program. Constellation spent several multiples of this number, well over $50 billion, on its launch vehicle and crew vehicle system and only successfully launched one suborbital rocket. Only today is the gigantic $30B SLS rocket ready to venture to the Moon.

NASA sponsored SpaceX Falcon 9 launch from Cape Canaveral carrying the first US astronauts on a privately developed launch vehicle. Image Credit NASA

The SpaceX experience is, in many ways, a model for how I see the next decade unfolding. "New space," as some call it, represents the hopes, ingenuity, and capital of investors to do what formerly was considered the sole domain of governments. Companies such as Skybox Imaging (later Planet), ICEYE, York Space Systems, and Iridium Communications are all shining examples of what can be done. For me, the future will be led by those willing to take risks and put skin in the game like the "new space" companies are. In the meantime, the government R&D community is taking it on the chin, and the World War II industrial relationship is going into the garbage bin of history.

The Inflection Point

My cell phone rang late on a Friday in July 2001. This was a pivotal year, and I was about to have a crucial phone call. I was unaware of how important these elements, the period of history or the phone call, would become. All of this came before the mind-numbing events of that beautiful September morning in the same year when our peaceful world was shattered by the reality that our enemies lived among us and meant serious harm. Before these tragic events, the world held a sense of optimism and calm that is certainly harder to spot ever since. Somehow the impossible seemed more plausible before September 11, 2001, and crazy ideas and thoughts a bit less penitent. Maybe it was the internet-fueled technology craze in full swing or the money and glow from that new promising technology still with us. Perhaps it was ten years of relative peace in the world and the absence of a threatening Soviet Union. Whatever it was, it seemed normal then, and so did the phone call I received. Oddly enough, the caller was driven by apocalyptic visions of humanity's end at the hands of larger galactic forces. While he seemed a bit out of character for the times, a few weeks later, he was to be proven tragically correct.

The caller on the other end of the line had an unusual accent, faintly British but not entirely. I have traveled all over the world and have seemingly encountered every variety of accents, but I could not place this one. The caller claimed to be an 'internet billionaire' and had vague notions about space missions using private funding. The private funding would be his net worth. He called me because he was told I was the right person to talk to who knew about Mars missions and Russian rockets. This was, oddly enough, true.

I could have never predicted the drama about to unfold before my eyes. I was no stranger to using private funding for non-commercial space missions as private citizens had funded me to work on a joint French-Soviet mission to Mars in the 1980s. I later worked with The Planetary Society to fly the world's first Solar Sail on a Russian Submarine Launched Ballistic Missile just the year before. This stranger on the other end of the phone had found me like a proverbial "needle in the haystack" out of an immense sea of humanity, hoping that we had similar interests and that I could assist him.

This 'stranger' on the phone was no less than Elon Musk. At the time, it seemed like a typical phone call. Someone with big dreams of flying missions in space and little experience at doing it, who was willing to pay to learn the ropes. The phone call from Elon reminded me of many of the wealthy characters I occasionally meet at the racetrack who took an interest in road racing. They would show up at the racetrack to absorb the energy of the undertaking and imagine themselves competing in fender-to-fender competition with the other drivers. Some get interested enough to approach other racers about learning to race. They always seem awkward and out of place in the beginning and somehow vulnerable. We racers are a small and arrogant community that smells newcomers from a mile away, much like the space industry. Some wealthy would-be-racers would try it out only to find that this was hard and gritty work with only a questionable return in pleasure and satisfaction. You must genuinely love the sport to continue throwing massive money, energy, and time at the endeavor as required to be successful at it. As a result of this combination of factors, few wealthy would-be-racers become good enough or interested enough to continue the sport beyond the curiosity phase. Very few become good enough at the sport to become competitive professionals like Paul Newman, Steve McQueen, and Pat Dempsey. Elon was this same guy, but this was the space business. In this same sense, the space business is the same as racing, except space machinery is more expensive by orders of magnitude, and our small clique of arrogant insiders are generally better educated and have cleaner fingernails.

Elon's ideas were big. They were enormous. Initially, I had no idea how big his ideas were, as he was deliberately modest in his explanation of intent. His initial approach to me involved 'doing something that would prove mankind could become a multi-planetary species.' Somehow this involved something going to Mars, and anything going to Mars needed big rockets, which were very expensive in the Western world. This was still when human space flight was the sole domain of nation-states, and the launchers were direct descendants of ICBMs. However, big ideas were nothing new, and many successful internet entrepreneurs were coming out of the woodwork with ideas of making money in space or simply making a mark in history.

I have met a lot of people with big ideas over the years, and some of them even had the money to make these big ideas happen. As was typical of the times, I often listened politely to big ideas with a modest skepticism hidden deeply in my pocket. It's one of those situations where you are constantly 'damned if you do and damned if you don't.' Some ideas are completely crazy, and a self-respecting engineer or scientist would not want to be associated with them. Yet other ideas are modest at first glance but have an underlying truth and brilliance that makes the idea more significant with time. And then there are those ideas that are so audacious that they take your breath away. Elon's ideas were of the latter variety, as time has shown.

Elon was different. You could sense it just from how he framed the problem and logically broke down the solution from first principles. I was not used to this thinking coming from the internet world. More typically, these folks would have a pet idea that they had been nurturing for many years and secretly wanted to pursue instead of the one that was making them money. In many cases, there was a good reason that these ideas were being hidden away from public view. Elon's first mission concept smelled a bit like this. Somewhere, he picked up the idea of flying mice on a round trip to Mars to demonstrate that they could survive the zero-gravity environment and breed. Technically the idea was sound. There was little doubt that mice could breed anywhere, even in zero gravity. Elon thought this mission would inspire humanity so that, by proxy, humanity could make this trip and survive. Those of us forming the technical team were less than convinced that this mission was either a good idea or, in the worst case, might lead to some 'humanitarian' rodent disaster in space.

Besides the technical problems, the public was long ago sold on the idea that humans could spend a lifetime traveling through space from one new world to the next. Star Trek, 2001: A Space Odyssey, Star Wars, The Martian Chronicles, and countless other works of fiction brought this idea into the lives of millions and millions of everyday people and didn't need to be sold. Sending mice on a two-way pornographic journey to Mars and back would do little to inspire humanity, and most of the early team recommended against this idea. Those of us advising Elon in his early efforts believed that people would respond to the idea that life could thrive on another world, specifically Mars. That would resonate deeply with the exploration gene buried deep within our DNA and create the potential for others to follow, thus creating a mass movement to explore and settle on Mars. That was, by the way, no minor goal. We came up with the idea of growing a lonely plant on Mars in a greenhouse fueled by alien water, Earth-based seeds, Martian carbon dioxide, and sunlight. It was an idea that excited the mind and told the story of what humans would do if they ever settled on the Red Planet. They would have to do as their ancestors have done for millennia, develop an agrarian society, and learn to "live off of the land."

Our ideas and efforts at the time seemed inconsequential and appeared to those of us who had been in the business as just one more idea that would slowly disappear into history's dustbins. We needed to understand what this effort would lead to and what bowling pins of history would tumble as a result. It's tough to overstate how insignificant these events seemed to us then. We all understood Elon was wealthy and had plenty of intelligence, motivation, and experience building companies. We all ultimately failed to appreciate how significant this man was and how this would affect human exploration's future.

The brilliant people surrounding Elon never even considered that he might eventually be the man sending the first humans to Mars, as I now consider very likely. Despite this, Elon remained quietly and supremely confident in his vision. I saw this early on and was trying to figure out how to deal with it. First, there was the idea of sending the mice to Mars, which was bizarre and irrational. Even his good friend Adeo Ressi, whom he had met in college and had served as his informal consiglieri, thought this idea was a "bizarre philanthropic gesture" and even staged interventions with family and friends to prevent Elon from 'throwing his fortune away.' Despite all of that, Elon saw something that none of us saw, and there was little to deter this man from achieving his vision.

Who can imagine what might have happened had Steve Jobs and Steve Wozniak not become friends and collaborated on those early computers that eventually led to the technology that became Apple? Would someone have finally brought an equivalent technology, or would our world of instant communications and personal everything on our phones and iPods not exist? It's hard to say. Likewise, we might have faced a much darker world had Werner Heisenberg, leader of the World War II Nazi nuclear weapons program, not made an error in his early calculations regarding the neutron moderation qualities of graphite. His calculation error was a critical "fork in the road" that led the Nazi bomb project down a path that made the production of a German atomic bomb unattainable in the timeframe of the 1940s. On the other hand, the U.S. scientist Enrico Fermi made the correct calculation, leading him to the successful first nuclear reactor constructed on a handball court at the University of Chicago. History will never know if this confluence would have changed the destiny of mankind, but indeed it may have. When Elon called me and I gathered the troops in 2001, history was being made, and none of us appreciated the storm that it was to launch.

It has been said that a leaf dropping from a tree on a late afternoon in China provides the seed for an atmospheric disturbance that becomes a hurricane by the time it reaches the shores of North America. Much the same can be said of human events involving Elon Musk and our early years together. The confluence of people, ideas, and timing conspired to create history in ways few foresaw at the time. Commercial space had existed at the time for at least 20 years, if not more, but what we were doing with Elon in late 2001 was reigniting long-dormant energy in the aerospace world fueled by people restless to get humanity moving towards the stars and fulfilling that unfulfilled promise left by the American Apollo program going to the Moon. The time was right, the seeds had been planted, and Elon was the new leader of this Revolution in space travel that would take the center of gravity in space exploration from the hands of nation-states and place it firmly in the hands of societies entrepreneurs and innovators. Thus, the quiet 'Space Revolution' began in 2001.

Those of us who participated in the Mars Oasis – or the "Life to Mars Foundation" as we called it then, had no idea that we were the first bullet in a Revolution that would eventually sweep our world and perhaps humanity later with it. Revolutions are not methodical, nor are they planned. Revolutions are not the result of continuous change or progress. Revolutions are unintended consequences of apparently mundane actions of individuals and groups, seemingly ordinary inventions, and unimagined discoveries. Viewed from the perspective of decades and millennia of history, revolutions appear as a sudden change in human direction, yet they are, at their roots, relatively minor activities and coincidences that combine to create a more significant meaning and a more considerable impact with time. Fundamental transformations begin with slow, unnoticed changes that eventually become torrents of uncontrollable alterations of reality. At times, it seems that change had a life of its own.

Revolutions are also begun by rogue individuals who are malcontents and unwilling to accept the order of things as they find it. The word Revolution conjures many notions and ideas in the mind. It can manifest in many forms, such as political revolutions, technological shifts, sociological changes, and even human thinking. All true revolutions fundamentally change the way humans see themselves concerning each other and the world that they inhabit. Once the Revolution is upon us, we accept it as the new normal and perhaps even acclimate to the fundamental shifts in the world around us. Such is the capacity for humans to yearn for normalcy and tranquility. But revolutions do happen, and they change the very course of history. And with most revolutions, they are acted out by unsuspecting players in unnoticed roles. The heroes and heroines of revolutions are ordinary people finding themselves in extraordinary circumstances doing ordinary things.

Elon is a revolutionary in the truest sense of the word. He instinctively gathered other rogue personalities around him to accomplish his significant goals. He knew he needed people like this to realize his vast dreams. Most of us that ran with Elon in those early days were of his mindset. Still, we lacked his supreme confidence and unwillingness to consider that any of this could ever fail if we applied the best of ourselves without reservation to the problem at hand. That's a big bet to make with your life. If you are a revolutionary, you are born that way and can do little to change that. All you can do is follow your instincts and hope that your instinct serves you well in your adventure through life.

It's been over 20 years since the first phone call from Elon, yet it seems like yesterday. Time and distance have allowed me to reflect on the real meaning of the events launched by that call and to place them into a larger context. The larger context is important to understand how Elon became a transformational force in space exploration and humanity's place in the cosmos.

This body of work chronicles the events that led to this unique part of history. Elon and SpaceX didn't happen in a vacuum. Instead, Elon found and drew upon others in the space business. These individuals shared his discontent with the status quo. These elements came at the right time and place to spark a revolution. We didn't think of it as a revolution at the time. Rather, it was returning nation-state space leadership to the people. We didn't do this by confronting the nation-states head-on in a challenge. Instead, we went ahead without them. The resulting commercial space progress "sucked the oxygen out of the room," as the phrase goes. Elon Musk and other billionaires joined the Revolution as it was well underway. Their entrepreneurial savvy and immense wealth accelerated the Revolution. This is the story of the new space race, and my involvement with it.

Long live the Revolution!

CHAPTER 2

THE BUILDER

I have always been a builder, a maker of things. I am happiest when I am building something. This has been 100% true throughout my life. However, since my earliest childhood memories, I felt uneasy about my place in the world. Looking back on it from a perspective 50 years later, it was more of uneasiness with the status quo, with the way others saw the world around them, with refusing to take risks, and with staying in one place for very long. For me, the voice that called on my inner self to venture out into the unknown began softly at first as a child but eventually emerged as a gigantic decades-long passion later in life. Maybe it had something to do with the small dusty corner of the world where I grew up. Perhaps it was simply because I was never taught that I had bounds on what I could imagine, think, and do. In any case, I had to respond to this inner voice and move. Otherwise, I knew I would die.

My ancestors came to the shores of the New World in the early 1600s as explorers driven by a need to explore and make a new life in a place that few Europeans had ever even seen. My great-grandfather of many generations ago, William Cantrell, was an English gentleman and navigator who was one of the original Jamestown settlers in 1618. He arrived on the second resupply mission and later led an exploration party up the Potomac River to discover, among other places, the site of modern-day Washington, DC. His family later joined him in the New World and married other pioneer families from Jamestown. That DNA still lives deep inside of me and my children. It's a force that is undeniable and sometimes overpowers good judgment. Or so it seems.

It's hard for us to imagine today when we can see every part of the Earth simply by logging onto Google Earth and scanning our mouse across the screen. One can hardly imagine a world where we go somewhere, not knowing where we are going and not being able to know every detail of our destination. When my ancestors came to Jamestown, this was not long after the dark ages and was when science and reason were leading Western mankind out of a period of superstition and darkness. My ancestors heard that faint call of the open road and must have known it was their time to move.

No doubt that residual DNA burning deep inside me as a child fueled my notion that there must be something bigger out there in the world. This underlying sense made it hard for my friends to understand me and even harder for my extended relatives to comprehend my drive. This was my first sense as a child that I was somehow different from the others, and often not in a good way. I felt like a defective human being. Even my teachers occasionally treated me accordingly. I vividly recall one of my middle school teachers scolding me and telling me, ' You will be in prison before you reach age 18'. Perhaps she was right, but this was my first sense that, in my core, I was more like a cowboy at heart, at ease alone in the desert with my thoughts.

Ever since I was a very young child, I tormented my mother by tearing apart her household appliances to see how they worked. I would build something 'new and different' with these appliances' mysterious parts. My mother would sigh upon discovering yet another massacred appliance or electronic device, and patiently say, 'This is a creative mess.' My mother was a saint of a woman and withstood my immense curiosity and destructiveness with great patience and wisdom.

As a child, I grew up in an era where we still had relative freedom to come and go as we pleased without fear of harm. I grew up on a chicken farm where we had a can-do attitude, space, and resources to build things. This was the perfect place for a curious child. I spent days wholly immersed in the project du jour through summer and on as many school days as sunlight and time permitted. I would build anything ranging from go-carts, forts, and television antennas to mechanical dog feeders. I had a treasure trove of natural resources to draw upon: whatever I found lying around the barns and workshops on our chicken ranch. If anyone bought lumber and didn't consume it quickly, it was in grave danger of being put to a purpose other than what it was intended for. All under-utilized wood was immediately appropriated for use on soapbox derby cars so that my friends and I could race each other to our impending deaths. My thirst for raw building materials and household appliances was nearly as inexhaustible as my curiosity about how the world was built.

The California I grew up in the 1960s and 1970s was, so we were told, a 'golden place and time'. I was not born into ambitious surroundings, however. My hometown of Yucaipa was sleepy and quiet. It was a place that time seemed to have forgotten but hung on the edge of something immense and vital: the megalopolis of Los Angeles. It seemed like a place where dreams came to die. But for all I was concerned as a child, Los Angeles was a world away and a place where movies were made, and people spent hours parked on the freeways. Yucaipa quietly existed in the shadows of that immense energy and hustle. It was a nice place, a good place for a kid to grow up in. By the mid-1970s, Yucaipa had little in the way of crime, and a young person could roam free on a bike during a summer day visiting friends, collecting soda bottles to return to the store and get five cents per bottle, and to use that loot to buy candy or, better yet, implements for building something.

This was a time before constant connectivity, cell phones, and the world was so horrifically connected that our every movement could be known and monitored. If your parents were wise, they made you call and tell them where you were going before you left the house and report back when you returned. Finding a friend, or your errant child, for that matter, involved making phone calls to all the homes in the neighborhood to begin a search in a manner reminiscent of the best detectives you could ever imagine. It was a chore. Yet life was simple in this time and place and most of us who spent our hot summer afternoons building forts, building go-carts or motorcycles, and discovering the joys of being a child could ever imagine the world that would unfold in front of us 40 years into the future.

When I was a child, the Soviet Union was an existential threat to our very existence that seemed very far away. We would only know about it by watching the nightly news, reading the newspapers, or participating in the civil defense drills at school designed to prepare us to survive a nuclear attack. We, as children, could see through the apparent lack of logic in this exercise. We knew in our hearts that crouching beneath our desks would not prevent death from a nuclear weapon exploding in our vicinity. We grew up near Norton Air Force Base, where nuclear weapons were stored, and March Air Force Base, where the Strategic Air Command bombers were awaiting Armageddon. As kids, we found other amusements in the activity, like identifying what color underwear the girls wore that day or passing along secret hand signals among the cognoscente. Strangely, this place in time was caught between the fantasyland of white picket fence America and the cruel reality of a harsh world on the outside that genuinely meant us harm. This perspective never left me as I grew up and entered the world. I knew inside that I had grown up in a very peaceful place and that my youth was protected from a much harsher outside world.

My hometown of Yucaipa CA was your everyday small town USA as can be seen from this view of California Street. Lusby Motors is on the right. Image Credit Jim Cantrell

Despite its sleepy and agricultural nature, Yucaipa is at the crossroads of people and events and became symbolic for much of my later life. It was the kind of town where nothing remarkable or newsworthy ever happened, and nobody famous emerged. Yet, if one was patient and watched, one could see famous people pass through Yucaipa quietly between Palm Springs and Los Angeles. Occasionally fate would bring them into town for car repair or a meal. The locals knew this small fact and delighted in seeing the occasional famous movie star or politician in the stores or a local restaurant. Rumors occasionally circulated about John Wayne being sighted and overheard eating lunch here or other favorite movie starlets dining in sunglasses hoping not to be noticed by the public. Whenever someone famous passed through the town, word would spread widely and quickly about the sighting in a manner reminiscent of the best gossip pages.

Despite the apparent self-love of the locals, Yucaipa was a place mainly to pass through and offered little out of the ordinary to entice the passerby to remain and make a life. To the outsider, it was only notable for being a mile marker on Interstate 10 and a place to stop and get fuel on the long journey to the desert. If you were approaching Los Angeles from the east, Yucaipa greeted newcomers with the first sight of green grass and trees after a long desert trek and offered convenient rest stops before heading into the greater Los Angeles area. Orange groves, apple orchards, chicken ranches, and retirement communities were the primary industries in this small town, so typical of California in the 1960s and 1970s. It was the kind of small dusty corner of the world where time seemed to stand still and where a kid could grow up and not worry too much about encountering the complexities of life until you were ready to wander beyond its confines. It became, for me, a place that symbolized ordinary American life in the 1970s and a symbol of how extraordinary things sometimes emerge from very ordinary circumstances.

The 1970s world I grew up in was very much different from today. This was an era before computers, mobile communications, and the internet. Information was hard to come by, and much of what I learned about machinery and technology came from library books and magazines such as Popular Mechanics and Popular

Electronics. It's hard to imagine a time when you couldn't fire up a laptop computer or mobile phone and type in some terms into a search engine to drum up a quick Wikipedia summary of a topic of interest. The internet has yielded instant global intellectual democratization by making knowledge access much more universal. Knowledge is power and the power to make our destiny. As a kid, I would spend countless hours scanning magazines and lists of books for sale and wondering how many lawns I would have to mow before I could afford the book on stereo amplifiers or internal combustion engine designs. Imagine the disappointment and simultaneous thrill I felt when I finally spent my well-earned money on books I could hardly understand. But I had time on my side and plenty of it on those long summer days and endless summer nights. I would read and re-read these books, hoping to cultivate a seemingly secret language from them that would permit me to build ever more sophisticated machinery.

My childhood home was built on a hill and was imbued with a gift from God to a boy like me: a driveway that went around the house and down the hill. The driveway swept a 'figure nine' around the three homes. Here, my grandparents and family lived like roosting Italian chickens. The driveway had a gradual drop to it followed by a long straight section that was perfect for gaining speed. The steepest part of the driveway involved a 60-degree corner which, if improperly navigated by a budding race car driver, led to an unfortunate encounter with the oleander bushes lining the driveway. It was the perfect place to stage soapbox derby races. In later years, it served as an ideal racetrack for our homemade gasoline-fueled contraptions cobbled together from steel and wood that we called go-carts. The sense of accomplishment in imagining a machine in our minds, followed by days and weeks trying to figure out how to build it with the materials on hand and knowledge in our heads, was immense. The most exciting part was finally making it fly down the driveway; the combined thrill and adrenaline rush was hard to top. It kept us away from girls and drugs – both of which my parents were worried about. Our drug in those days was the smell of gasoline and the feel of the wind in our hair.

As a kid, my house was the natural magnet for other like-minded individuals who would flock to my parents' barn, which I had converted into a makeshift workshop. It attracted the neighborhood's motivated, energetic, and creative minds and filtered out (mainly) the TV watchers, the lazy, and the otherwise uninteresting kids in Yucaipa. I spent evenings when lighting conditions did not permit, working in the barn on new contraptions or riding them down the hill, designing new soapbox derbies which eventually graduated into steel go-carts and later automobiles. I drew my car designs, sent them to the local newspaper 'junior artist' contest, and won several awards. In many ways, this formed the fundamental basis for my adult life and career in the space industry. The go-carts and soapbox derbies were our 'start-up' businesses and received all our energy, creativity, and money. It was a childhood that everyone should experience.

The family compound in Yucaipa CA. Image Credit Jim Cantrell

My imagination and ambitions knew no bounds. It still does not today. As a 10-year-old child, I dreamed of building a majestic and immense theme park from the lumber at hand and building castles on the hillsides. Later, as my sophistication increased, I dreamed of building the Starship Enterprise. I started building my first theme park on our vast six-acre chicken ranch. After a few short days of that hot summer in 1976, I ran out of building materials and nails. I suddenly faced the reality that I would need additional capital to realize my theme park dream.

After visiting the local lumber yard, I prepared detailed drawings and cost estimates and presented this to the best venture capitalist I knew: my mother. She was patient and listened to my tale and passion for building this theme park. I even took her on a tour of the site where it would be built and outlined the massive revenue it would bring in. I thought she would see the plan's brilliance and invest her meager savings in my hair-brained scheme. She was kind in rejecting my business plan citing safety and insurance concerns rather than focusing on my lack of experience. I eventually dismantled the initial structures to salvage the lumber for more soapbox derby racers. If I couldn't become a theme park magnate at the age of 12, I would at least spend my days going fast and cheating death.

By far, my greatest obsession as a child was things that moved on four wheels. I liked anything that went fast. I tried to make things pictured in the books I had borrowed from the library, and I could use the tools at hand. For parts and supplies, I turned to abandoned lawn machinery. An unused lawnmower was immediately stripped of wheels for use on our soapbox derby cars. My father was none too patient with me and my lawnmower wheel fetish. The prize was the wheel and the speed that this revolutionary technology gave us. Any amount of my father's yelling and beating was indeed worth it. Nothing could deter me and my friends from the pleasure we got from riding a machine of our creation down that hill. Nothing else in life compared to this.

We built many variations of the soapbox derby, from small and fast to large and creepy. We had one model called the "Cadillac" because it could haul the whole gang down the hill, and the group could all push it back up the hill. It was very fast, and we all laid down to minimize aero drag and the potential decapitation from the wire fence lining the left side of the driveway. The main problem was stopping it at the bottom of the hill. The technology we developed for single-person machines, a single piece of wood levered against the ground acting as a friction anchor, was insufficient for the additional weight of the riders. After the first time we rode down the hill, we were all relieved that no cars were coming down the street. Unable to stop, we dashed across the street and into the adjacent gravel driveway. The "Cadillac" was deemed dangerous and chopped up to make other smaller soapbox derbies.

My soapbox derby days forever changed when I discovered a book in the local public library entitled 'Build Your Go-Cart at Home from Everyday Items.' How could I resist? Were there other people who thought like I did and were interested in these things? To her detriment, my mother added the book to the pile I was checking out at the Yucaipa Valley County Library. I was rapt for hours as I pored over its contents. It was simply brilliant. Take a new lawnmower, remove its wheels, and use its engine for propulsion. To an adult looking in from the outside, this was a terrible waste of a perfect lawnmower and a perilous way to risk a child's life. As for me, I envisioned wind in the hair, a practical vehicle that could help with chores around the house and a project that could sate my appetite for building something new and challenging. My parents either didn't know the depths of my plans or were surprisingly open-minded about the project. I honestly don't remember asking for permission.

It's probably good that my father was not very uptight about mowing the lawn on a fixed schedule and that the grass went dormant in the late summer heat. By the time he discovered that I had disassembled his only running lawnmower, used four sheets of plywood, and had a barely recognizable contraption to show for it, he might have killed me if it were not for the benevolence of my mother. She spared my life that day by calming my enraged father and taking him to Sears to buy a new mower. I heard the hushed conversation through the bedroom door with my mother using words like "curious" and "creative" and phrases like "it's a creative destruction." For my father, a child of the Great Depression, this could never be OK.

I could not understand my father's perspective at all. After all, he also loved fast cars and mechanical contraptions. How could he not support this, I wondered? Life continued, and I eventually got the go-cart to move under its own power. That was one of the most exhilarating moments of my life when the engine fired up, and I sat watching the little single-cylinder motor chugging away on its own.

I had yet to pilot the wooden contraption, but it was almost enough to see it living and breathing. The machine must have had magic because it seemed to be a living and breathing thing in my mind. That magic spark has stayed with me throughout my life and has been the continual waypoint in this adventure that tells me I am doing something right. Ultimately, this first wooden vehicle was not the most reliable machine in the world and was not faster than the now-successful soapbox derbies. Undeterred, I vowed to never, ever, ever give up. My opportunity came later that fall.

On my way home from school one day, I noticed a metal frame go-cart in front of a house. It had been sitting there for a long time by the caked-on dust and flat tires. It looked sad and seemingly cried out to me as I rode by every day on my bicycle. I imagined that it spoke quietly into my ear each time I rode past, "Please come help me, love me, and take me places with you." I rode past that house daily and watched to see if it had moved. Finally, one afternoon, I got brave and knocked on the house's front door. A lovely lady answered the door, and I explained my interest in the machine taking up valuable space in her flower bed. I tried my best to sell her on the idea that this machine was in her way and that it would be much happier being somewhere else, like with me. I imagined she would be as interested in seeing it gone as I was in riding it around my racetrack just down the street. She took my phone number and name after explaining that it was her son's go-cart, that he didn't use it anymore, and that he might like to sell it. I never mentioned it to my mother as I thought that the owner of this fine machine would never want to sell something so wonderful.

The call came to my home several weeks later, and fortunately, my mother answered the phone. I had forgotten entirely about the contraption. In one of the most memorable moments of my life, my mother and father bought the go-cart for me for twenty-five dollars and brought it home in the back of our beat-up brown 1963 Chevrolet truck. When we got it home, my dad got it to run. It was a magical moment that I would never forget. Sparks were flying from the exhaust as the engine revved and coughed out the various layers of dust and debris from the engine. My father was even happy as he realized that his new lawnmower would be safer this way. When I drove the new contraption in the dark for the first time, I had a broad smile that took days to go away. I was hooked. My life was forever changed from that point forward, and I knew in my heart that the world was no longer the same place it was several short minutes ago.

Part of the price I paid as a child for being a builder was not having many friends. To be sure, I had lots of acquaintances who wanted to ride the go-carts down the hill with me, but they were friends of the hour or, in some cases, of the minute. The truth is that I was a loner, and I was okay with that. Being a loner meant I had more time to build things, daydream, and think. As a child, I truly enjoyed the freedom to think and imagine, but the price was social isolation. Some of the kids in the neighborhood didn't understand me very well and made fun of my weight. I was a very short and plump kid and was mercilessly teased by the others and ostracized for being large and clumsy. I was far from being athletic as a child and was always the last kid to be picked for the baseball team. My other friends called me a 'brain,' which further drove me into my isolation. I spent my summers creating, building, listening to, and writing music while not building or driving new go-carts. I learned how to play the piano and began composing my tunes. It was a welcome relief from the terrible teasing from my friends in my late pre-teen years. This situation was compounded by my father, who became an alcoholic after the death of my sister Vickie. As the family disintegrated around me, I had my machines, my music, and my imagination to comfort me, and that was enough.

The awful teasing lasted until about my 12th birthday when I added about 10 inches to my height in over a year. This growth spurt made my once pudgy physique into a slender and muscular young man with a large and sturdy frame thanks to my continuous physical activity associated with my building and working. All in all, this was an excellent thing and served to help my confidence. It also helped that I became a giant young man at school and was suddenly feared instead of teased. I am not sure if the fear came from the sudden increase in size, my quiet demeanor (which was caused by wanting to avoid being teased), or the fact that I beat the hell out of one kid who, on the first day of school, asked me where I "put all that fat"? I was angrier than determined to send a message to this boy, but between my newfound strength, size, and budding confidence, my days of being teased by other boys and girls were over.

As my early teen years passed and I grew older, my projects became more and more ambitious, consuming more time in the planning stages, requiring more research, and of course, requiring more money. To me, it was a shame that money became my limiting factor. To my parents, a lack of money was of great solace as it slowed me down and enabled them to throttle my expansive energy. Because of this never-ending appetite to build and needing money to fuel the projects, I discovered the ability to earn money at a young age. In 1977, my parents helped my sister and her husband buy a small local automotive repair shop and towing agency. I will never forget my first tour of the place located on California Blvd in Yucaipa, California, oddly named "Lusby Motors" after its first owner Seth Lusby. It was indeed a "downtown" kind of place to my young mind, and it was a place where a lot of things were seemingly happening all the time. I found this energy intoxicating, and the machines being worked on to be what dreams were made of. I begged my brother-in-law for a job as I desperately wanted to be a part of the action and make money to fuel my off-hours projects. He eventually relented and allowed me to spend summer days cleaning floors, cleaning the various tools and machinery, and eventually working on the cars. I started slowly by changing the oil on cars and moving my way to changing brake pads. In my mind, I had become a 'big-time mechanic' at the age of thirteen, and this new job provided me with both income and satisfaction with my desire to make machines.

I wasn't making machines yet, but I knew it was only a matter of time. One of the wonderful side effects of this job was the people I encountered. People would come into the shop to make repairs or drop off parts. I learned about this secret underworld of people who maintained the machines everyone needed to live. These people were like demigods to a boy like me at thirteen. One of the guys I met was a man who ran Joe's Machine Shop and Welding. I called him Smoky Joe because he constantly had a cigarette hanging from his mouth while he worked, talked, or complained about anything that bothered him. Joe was in his mid-forties, slightly overweight on top of his 6-foot-tall frame, and strong as a horse. Joe's hands were permanently stained by the grease and oils from the steel and aluminum he handled every day, and this matched his deeply salt and peppered beard that he wore proudly without ever trimming it. Joe was a magician with anything metallic, and I would stop by anytime I was in the neighborhood. I was always begging for scrap steel from him for my go-cart projects. At first, I would give him whatever change was in my pockets to pay for the metal, but he eventually told me to keep the money. He liked me and called me an "enterprising young lad." He took me under his wing and showed me how to weld steel and some basic metalworking techniques. This was pure magic to a young teenager and opened the door to building many new things. I felt like I had been the apprentice to the Wizard who dwelled in the greasy corners of the blacksmith's shop. He taught me the sacred secrets of fusing metal pieces to form something new and more robust. If this wasn't magic, then nothing else was. Joe didn't know it then, but the gentle attention and kindness he showed to a young kid he barely knew would be remembered 40 years later, long after he was gone.

Early go-cart built for pleasure and racing.
Image Credit Jim Cantrell

I continued working at this job well into my early college years, and Lusby Motors became a core in forming what I was to do later in life. My first car came from Lusby, who had taken a 1955 Chevy in on tow, and the owner never paid the towing bill. I got this car running with little effort or time and, within three weeks, had it sold to someone in town who wanted it. At the age of 14, I bought and sold my first car. This became a lifelong habit of mine, but the sweetness of the first sale remained with me to this day.

At times, I am sure that my relatives must have thought I was born from some alien encounter of the fourth kind as I could not have been more different from my father and many others in my family. I had two sisters who shared little of my intense interest in machinery, electronics, or anything technological. My father delivered uniforms for a living and was happy to return home after a hard day and drink beer rather than challenge himself with something like an 'unsolvable' problem. By contrast, I spent many summers in my bedroom reading the Encyclopedia Britannica (which I imagine my mother was conned into buying by a door-to-door encyclopedia salesman) on whatever topic piqued my interest that day. After a hard day at work managing a nursing home, my mother would return home only to be redirected by my curious mind and insatiable appetite for information to the local library for books on the exciting topic of the day. Sometimes she would have to stop by Radio Shack to buy the latest parts for some crazy project I was cooking up. I am sure that she must have hated that store. It was with great sorrow that I watched recently as the last Radio Shack store closed in my Arizona neighborhood, knowing that an era in America passed along with an era in my life.

I did not understand my family. I am sure that my inability to understand them was matched every bit as much as their inability to understand me. This caused the distance between us to only increase with the passing years. The same was true of extended family members, few of whom had college educations or an appreciation for anything but the blue-collar work ethic. What entertained and fascinated me seemed to be a waste of energy and resources to them. I eventually ended up feeling disconnected from the blue-collar world of my youth. But coming from this blue-collar world, I never entirely became a part of the white-collar world I was gradually entering. My blue-collar roots often showed through my rough mannerisms, salty and coarse language, and continued enthusiasm for getting dirty. I never felt at home in the white-collar world I was slowly becoming a part of or like the people that inhabited it with their polite manners, clean fingernails, and clean language. Somehow, I felt disconnected from them, like I felt disconnected from my blue-collar roots and the group of childhood friends that teased and shunned me. My family and friends in my blue-collar world watched me do things that they scarcely understood and often attributed to a desire on my part to 'appear sophisticated.' While this was far from reality, I am sure these thoughts made complete sense to them, as no sane child would be doing the things I was attempting to do at such a young age. Now and then, however, I would come across someone different from this blue-collar world who also held intellectual curiosity as I did and, more importantly, an appreciation and passion for building machines. These people formed the core of my reality as I transitioned from a young teen into a young adult.

At 14, I decided to design and make my stereo amplifier with a then Earth-shattering 50 Watts per channel! This was a plan born of necessity as I loved hard rock music, and the louder it was, the more I loved it. On the occasions that took me anywhere near a music store, I would spend as much time as I could looking at the costly amplifiers and had most of the prices memorized. The simple fact was that I could, and did, engineer my speakers and build them to avoid paying the 1000 dollars that a good pair commanded, but I needed help to afford a high-powered amplifier to match it. So instead of saving money to buy a new or used amplifier, I built one myself. Not only did I plan to build a stereo amplifier that represented the most powerful system in the world, but I also imagined building and selling these breakthrough stereo amplifiers as a business proposition. I even had a name for the company "Advanced Audio Concepts." Little did I realize that I would later meet a like-minded soul named Elon Musk, who did the same thing as a child but with a software company and a gaming arcade halfway around the world in South Africa.

To design the new amplifier, I started with a library book on amplifier circuits, magazine articles on how to make circuit boards, and a self-taught soldering technique. After reading just a few of these excellent knowledge and experience sources, I was in business. I also discovered mail-order catalogs and abused my mother's only credit card, repeatedly ordering inexplicable things like resistors, capacitors, transistors, and integrated circuits. Every day was like a new adventure for me when the mailman arrived. My mother did not pretend to understand what I was doing now. She understood the go-carts and even the theme park, but I was disappointed when she could not explain the nature of the electrical current flowing in the home's walls. I thought my parents knew everything, and I was subsequently disappointed to find the limits of my mother's intellectual capacity.

My mother was a saint once again, spending countless hours transporting me to Radio Shack to support my dream of the stereo amplifier. She even offered to buy parts from catalog order houses and electronics kits from other outlets. For my part, it was a completely new world from the go-cart and opened the possibility of building things that I would never have even imagined. I got proficient at designing and building my own circuit boards and managed to store the dangerous etching chemicals without creating a fire or giant holes in my forearms. My family used to utter, "God looks out for fools, drunks, and babies," perhaps I was two of the three. Nonetheless, after several years of toil, several capacitors blowing up and filling my bedroom with smoke, three complete redesigns, and hundreds of dollars financed by my mother (the reluctant patron saint of science), I finally made the blasted amplifier work. And work it did. It was so loud I could knock pictures off the wall with the bass from AC/DC songs and irritate the neighbors to the point where they summoned the police on a few occasions. I still have the amplifier today, and it still works. I am afraid of leaving it plugged in anymore, but occasionally it gets used for unusual purposes, like when my son used it for a guitar amplifier in the late 90s.

As I approached 13 years old, my taste in science started to mature, and I began watching the PBS documentaries like NOVA and local specials. I also watched the TV show COSMOS taking in all the physics and space travel discussions that Carl Sagan offered up every day. I was mainly obsessed with COSMOS. Carl meant it to be educational; to me, it was that and something more. Carl's ability to describe how the world worked and came together gave me a sense of awe and wonder. This was a man that understood me, and I understood him. We were the same alien family that felt the magic in science and science as a force of good and progress. It was magical, like speeding downhill in a wooden contraption of my design and manufacture. Carl showed me a mysterious world beyond my chicken ranch reality in 1960s and 1970s California.

For those that didn't experience the COSMOS series, it was nothing short of the most spectacular TV series of the 1970s. It was the first time that science met entertainment and presented science and scientific literacy in a way that engaged all levels of intellect and society. Carl Sagan, the show's host, was an immensely talented astrophysicist and an even more talented educator. His passion for science and the importance of science in the future of humanity was both contagious and inspiring. For the first time in my life, I saw science as something approachable and started understanding the place of our Earth and humans in the universe. These were significant realities for a 13-year-old kid watching a primitive color TV set. I could not wait to see the universe and learn about it all. Suddenly my backyard became much more extensive, and the walls of the dusty corner of the world crumbled. Science and engineering were my way out.

The year 1978 brought changes to my life that would have lasting positive and negative effects. My father became increasingly alcoholic and often violent when he drank every evening and all day on the weekends. He only stopped drinking his cheap Buckhorn label beer when he passed out. I tried to ignore this man, but he found it hard to ignore me. During the summer of 1978, he started to get violent with me. Naturally, a young boy will endure the abuse at first, but by the time I reached 13 years old and was nearly six-foot-tall, something had started to change inside of me. I was truly tired of being bullied at school and home.

I resorted to drugs for a while to try and avoid the pain of the situation, but I found my moment one winter afternoon when my father asked me to do something that I thought was unnecessary. An argument ensued, and he set out to "teach respect" with his right hand in his enraged drunken state. It was easy to grab his hand, twist him around, bend him over, and stare him in his face while he was entirely under my physical control. He could not release himself from this lock as I was about 6 inches taller than him, much stronger, and about 40 pounds heavier. He and I locked eyes for what seemed an eternity.

My mother heard the yelling and came out to investigate. She found the two of us and began to take command of the situation in the most chilling of voices I have ever heard come from her. As she explained to me later, the truth of the matter was that she feared for his life. It must have shown in my eyes. I released him from my death grip, and nothing further happened as we both backed off. But I had witnessed that Sunday afternoon the death of something big. It happened swiftly, and it came without the anticipation or wish of either of us. Our relationship would never recover, and neither would my parents' marriage.

My parent's divorce was swift and without much in the way of discussion. It was never clear how we would get by, but we had a lot of family living on the hill like roosting Italian chickens, and somehow that seemed to make everything else bad that might happen less likely. My paternal grandmother had moved away several years before, and my sister and her husband moved into the house, continuing the line of succession of Italian Roosting Chickens in the family. Throughout the spring, my mother began dating an old family friend, Howard, and it was clear they would get married. By late June, they were married, and my mother had sold the chicken ranch. I would be leaving for what people called "Silicon Valley."

Leaving Yucaipa was not easy for me as I had some terrific friends there that I had bonded with through our mutual isolation and what I would later discover was our unspoken similar home lives. Most of them, too, had alcoholic fathers, and this fact was like a dagger in my heart when I discovered it in later years. Our social isolation came from our home environments, not from who we were. That was a profound realization that took me several decades to make. Despite the depressing prospect of leaving all I knew, there was much to see in the world, and I was determined to make the best of it. I headed off to Silicon Valley for a new phase of my life, determined to show the world that I could do great things.

CHAPTER 3

ON THE MOVE

"For all its material advantages, the sedentary life has
left us edgy, unfulfilled. Even after 400 generations
in villages and cities, we haven't forgotten."
- Carl Sagan, The Pale Blue Dot

I found it depressing when we moved from Yucaipa to San Jose,
located in today's Silicon Valley. I was now stuck in a tract
home that had few places to tear things apart, create moving
machinery, or do anything worthwhile. I had left my childhood
friends behind, along with my teenage sweetheart. It was simply
an awful time. However, my mother had remarried a remarkable
man that brought a whole new perspective to me and my life. My
new stepfather, Howard, was a good man and Vice President of a
semiconductor company in Silicon Valley. In the later '70s, Silicon
Valley was just that – based on silicon. The semiconductor industry
was king, and the technology rage then was the impending digital
age that seemed to take forever. It was a different and slower time.
The ties were wider, the bell bottoms more sweeping, the hair longer,
and smoking cigarettes was something seemingly everyone did. This
simpler Silicon Valley is to be missed and envied in many ways. These
early Titans of Tech created the foundations for a new machine that
would change my life - the computer. Being young, I could not see
where this was going, but this move to San Jose began a series of what
I now call "tripping over billion-dollar fortunes." My contact with

34

very extraordinary people who came to seem somewhat normal to a young teenager.

San Jose in the late 1970s was a sleepy place, but there was still an underlying energy that was hard to put your finger on. At the time, I didn't know that this would be the birthplace of the modern tech world and that giants like Apple were being built in my backyard. San Jose still had that sleepy feel of a transforming farming community stuck between a vineyard and a subdivision. It had its cultural differences too. Instead of the small-town atmosphere of Yucaipa of a near monochromatic racial makeup, San Jose offered a virtual melting pot of cultures, languages, skin colors, and foods from all over the world. The truth was that I became a racial minority in this new place, and it was both terrifying and a learning experience for me. Without realizing it at the time, this feeling of being outwardly different was good practice for the future. After a period of adjustment, I dealt well with the idea of being a minority and not always being treated well. It challenged me to rise above being part of a 'pack' and being like 'everyone else'. It was strangely refreshing and mirrored my early childhood isolation. I looked different from my friends, neighbors, and relatives on the inside, and here I looked different on the outside. There was much to learn from this new environment and other cultures and ideas.

In this strange new environment, I focused again on my first true love: anything with wheels. I was approaching the magic age of 16, which meant I was about to be imbued with a ticket to automotive freedom. I didn't have a car and would never be the guy that borrowed my parents' cars. Those were both boring and very uncool. I began working for my sister and brother-in-law at their auto repair shop in Southern California. I spent several summers working in the hot weather repairing cars and putting scars on my hands and body. Working on cars was hard and filthy work. Yet there was something deeply satisfying about working on these machines, and that feeling bordered on magic. It was an exciting combination of revealed mystery and a promise of adventure. Most everyone else there looked upon the work as drudgery and a way to pay for the weekend beer. There were those exceptions - intelligent men still found satisfaction and a living in this world of men and machines. One of these mechanic friends once made a very memorable statement that resonated with me regarding the adventure and wonder that came from learning new things. He had just finished pulling a car's dashboard apart, and upon finding the problem, he stood up and exclaimed, "I am going to need to drink a San Miguel after that adventure," which mocked a similar phrase from a famous TV commercial at the time extolling the adventurous qualities of the beer. I could only smile and laugh with my friend, who was old enough to be my father. Some of us found a common bond in this world that was hard to describe to those who didn't inhabit it. We knew who we were, and we had a silent but sacred bond.

I learned quickly from the other mechanics at Lusby Motors on California Blvd. They were patient and started with the basics of changing the oil and moved along quickly to changing brakes and minor repairs such as fuel pumps. My experience building go-carts also paid off as I continued my adventure into vehicle fuel systems and engines. I had learned the hard way by stealing my father's lawnmower engines that their carburetors were finicky, and the engine would often become hard to start. We knew little of what we were doing at age ten, but an entire summer offered enough time to fiddle and experiment with the machine to make it run. Most of the time, the carburetor was at the root of the problem, and I would figure out how to fix it. This skill impressed my new coworkers, and before very long, I was rebuilding complicated carburetors on cars and trucks and making more money. I soon learned that the shop charged its customers per job regardless of how long it took to finish it (so-called flat rate as is standard in the auto repair business), and I soon negotiated a flat rate fee for myself. This was one of my first lessons in the economic power of supply and demand. By the end of my first summer, I had earned nearly ten thousand dollars, a hefty sum for someone in 1980. I used that money to buy and fix up my first car.

My brother-in-law and sister Debbie ran a towing agency in addition to the repair shop, and one fed the other. I often went out on tow calls, and sometimes it was an experience in tragedy or fortune, depending on how one looked at it. Most calls were for dead batteries, cars that would not start, or the occasional stranded motorist. The calls that came through in the dead of the night were the most unusual and occasionally frightening. Often, it was a violent wreck with deaths, and we had to rush to the accident scene. When you arrive at the location, you are greeted by complete chaos. Emergency vehicles and crews ran around the accident scene, tending to the living and occasionally making life-and-death decisions on who to treat on the spot and who not to tend to. It was a definite adventure for a 16-year-old kid who had never seen anything approaching death. Often, these accidents reminded me of war movie scenes with the chaos, blood, death, and the intensity of the officials trying to save lives and eventually get the highway open. Only these movies were playing out and in real time before our eyes.

Our job as tow truck operators was to help the emergency personnel to clear the scene and clean it up. We had to stay alert and out of the way until they were ready for us. This left me with a lot of time to ponder the fate of those before me and, by extension, my fate. In total, I must have seen a dozen people dead or die before my eyes in traffic accidents, and while it never seemed normal, it brought me to terms very quickly with my mortality. We could never seem to get over the death of a child. I only saw one, but I will never forget the sight of a grown California Highway Patrolman openly weeping as the EMT removed his lifeless body from the car. Maybe it was the indescribable sorrow of a parent who has lost a child or the realization that this human being could never get a fair shot at life. Either way, it was a sight and sound that stayed with me forever and confronted me with a reality I had never forgotten.

Of the many different middle-of-the-night tow truck experiences, one incident remains with me. It began with a phone call from the California Highway Patrol dispatch center in the dead of night. Phone calls arriving in the middle of the night when good folks are sleeping rarely bring good news. This one was not an exception. The CHP would rotate phone calls between various towing agencies as incidents occurred; this time, it was our turn. My brother-in-law answered the phone and got the message that there was a double fatality accident "on the mountain" near Big Bear City, high in the San Bernardino mountains. Waking up at two in the morning and getting into a cold tow truck was always a surreal experience. There

37

was little radio at 0200 in the morning as radio stations used to take the night off the air. This was in the 1970s, well before satellite radio and 24/7 media buzz ruled our every minute, and the drive usually took place in blessed silence. I got used to listening to the mechanical noises of the tow truck from the tire knobs as they crossed the road beneath me and to the low-frequency whine of the gearbox and rear end punctuated by the ever-present chain clanking on the rear body of the truck. This kind of silence left you alone with your thoughts. I found it somehow meditative, which often became necessary for the carnage that awaited.

My brother-in-law and I arrived at the scene an hour later after a long and methodical drive up the mountain. We knew we were approaching the scene as the floating multicolored lights among the trees gave the location away many miles before we arrived. The multiple colors of red and blue danced across the summer night and through the dense pine trees in a rather persistent but random way. Rounding a long sweeping corner, we slowed the tow truck as we saw at least four highway patrol cars parked on the side of the road with officers standing around talking to each other. Oddly, there was no car or accident in sight. Typically, such scenes offered up a gruesome sight of mangled machines and emergency response teams rapidly moving about. In this case, the emergency vehicles consisted of a single fire engine parked on the side of the road accompanied by an ambulance. This accident was undoubtedly a fatality, but we didn't see anything that would otherwise lead us to this conclusion.

As we stepped from the relative warmth of the tow truck into the cold mountain air, we could hear the chatter and laughter of the patrolman, but it began to quiet as we approached. We were greeted by one of the patrolmen who asked if we were ready to go to work. In response to our puzzled looks, he depressed the heavy button on his big black flashlight that doubled as a kinetic weapon and traced an arc from the ground, down the hillside, to the bottom of a pine tree, up the tree some forty feet and stopped on what looked like a car wrapped around the trunk of the pine tree. His light dwelled on the object in the tree that looked like a hot towel wrapped on a telephone pole. I could see that it was a car because the entire drivetrain was exposed on the outside, and you could see the complete undercarriage of the car. In a somber voice, the patrolman said, "There are two deceased souls in that car. An EMT climbed up there to confirm no pulse in the two passengers. We need you to extract the car from the tree somehow so we can remove the bodies from the car with the jaws of life. Oh, the fuel seems to have leaked out and evaporated, so we think you are safe for fire". I could tell right then that this was going to be another of those life-altering experiences for me.

My brother-in-law was a man of few words. He nodded at the patrolman, took the flashlight, and motioned me to follow him down the hill as we got a closer look at the tree and the car. It was not easy to climb the tree because of the density of the branches, but at least they were dense and horizontal. We looked at it for a few minutes and decided that the best thing was to climb up the tree on the side of the trunk that would not likely see the car fall should it do so while I was climbing it. Elroy returned and moved the tow truck to the edge of the road closest to the tree and unfurled the large horizontal stabilizers in anticipation of pulling the car off the tree. I walked up the hill and grabbed the hook and winch cable, and let it direct me down the hill again as it spooled out of the winch. Letting enough go by the tree, I motioned to stop the cable payout and then attached the hook to my pants. I climbed the tree slowly, trying not to move it as I was deathly afraid of the car falling on me. I reached the car and could smell the stench of death. It's a hard smell to describe, but it was a distinct odor of bodily fluids mixed with vehicle lubricants and antifreeze. As I tried my hardest not to inhale for fear of inhaling death, I managed to tie the hook to the car's rear axle. I could tell by now that the car was a Chevrolet Vega - an economy car from the 1970s and I could see now drying blood as it was dripping from the car. It was a sickening scene, and I couldn't wait to escape it.

I climbed rapidly down the tree, not considering my safety anymore. I wanted to get as far away from misfortune and death as possible. I must have looked like a monkey climbing down the tree as the patrolmen chuckled at my apparent hurry. I hurried up the hill and said nothing as I sat beside the truck. I could hear the winch pull in and the strain on the electric winch motor as it began tensioning the cable. After a few seconds, the truck started to move as the heavy winch cable started moving the doomed vehicle and passengers off the tree. I got up and walked to the patrolmen huddled near the tow truck. They had multiple big heavy flashlights trained on the vehicle when it came loose and fell 30 feet to the ground. It landed with a quiet thud, followed quickly by the sound of several firemen moving down the hill.

The car was a complete wreck. The force of the collision with the tree folded the car around the tree. It was apparent from the damage that the vehicle had left the road at a very high speed, made a pirouette about its roll axis, and pitched over ninety degrees to hit the tree directly on the car's top across both the driver and passenger seats. The two poor souls in the car had no hope of surviving. The impact had folded the car around the tree with such force that all four tires were at least a foot off the ground once on level ground. The emergency crew quickly began peeling the top off the car to recover the bodies for transport to the San Bernardino County Coroner's morgue. There was little question that they were both deceased. What was tragic about this scene was that the two kids in the car were brothers out for a test ride of their vehicle in the mountains. One was barely 17 years old, and the other 15 years old. They had put a huge engine in a tiny car and ran out of talent as they piloted this home-created monster up the mountain curves. I was told later that their mother was inconsolable as several of the patrolmen had the dubious task of informing her of her sons' deaths at an unholy morning hour.

This incident has stayed with me throughout my life and remains vivid. Steve Jobs was famous for saying that "one of life's best inventions is death because it reminds us of the finiteness of life." This incident embedded in my mind that my love of speed and fast machines was a deadly addiction and that I was not immortal. It also left me with the idea that I only had a finite number of years remaining on Earth, and I had best make good use of them.

My older sister had died some eight years earlier from a tragic encounter with Hepatitis, and this incident showed me that I was not immune from the same fate. There's no point in fearing death at every turn, but there is a tremendous point in understanding the risks of what you do. Every time I strap myself into a race car, even today, I am reminded of this event and that a demon is out there above 200 mph waiting to meet me. Every time I get in the car, it is my job to avoid that demon. To this day, I still have my superstitious ritual to prevent that meeting. I must place my wedding ring in the right pocket of my driving suit, the shoes go on in a specific sequence, the laces must be tied in a particular manner, the belts tightened in a pre-ordained sequence, and the mind must be placed into that peaceful place all us racers go before our encounter with the speed demon. This is a holy ceremony that most people never notice. For those of us who have seen the monster face to face, we religiously adhere to our little ceremony.

My annual return to San Jose from the summer working at Lusby Motors was always an event of sadness and change. I transitioned from an environment of constant action and adventure into a quiet, sedentary life. I could study, read, and build electronics, but the environment differed from the world of men and machines I had left behind. San Jose in the late 1970s and early 1980s was, in retrospect, an excellent place for a free-minded young man to grow up, and I slowly began to discover the inner freedom that such free-thinking afforded. There was much to see once you looked beneath the veneer of tract homes and endless freeways. The Stanford campus was in Palo Alto and had various summer programs for bright young minds. Santa Cruz had terrific beaches if you could stand surfing in the cold Pacific water. And San Francisco had anything a young mind could imagine and then some. Some of the most exciting people in the world can be found in San Francisco.

One of my most vivid memories of this period was the 'human jukebox' who played trumpet on Fisherman's Wharf in the 1970s and 1980s. The human jukebox was Grimes Poznikov. He would wait in a cardboard refrigerator box until a passerby offered him a donation and requested a song. I always remembered the rolled cloth front flap of the box that he would open and play a requested song on a trumpet. Occasionally he would play the song on a kazoo or clarinet. Some thought he was a lunatic. Others saw him as a free spirit. In many ways, he symbolized the free spirit of an age that was the fertile soil for a whole generation of technology innovators and rogue human characters.

The public school system crushed the creativity and character of the rogue thinkers. Many of us carried out our odd activities away from the view of schools and in hiding. The system did not encourage playing with computers, electronics, and machines. Instead, they valued the feel of paper, books, and classic education over working with their hands. As a result, I began carrying on what amounted to a double life. By day, I was a studious young man who studied mathematics, architecture, and history and played classical and jazz piano. By night, that world of men and machines I left behind tore at me to return. Doing well in school made my parents happy and much more tolerant of my increasingly ambitious mechanical aspirations and the money I spent on them.

I returned annually to work for my brother-in-law during the summer months, but I would build my first car this time in 1980. I picked a vehicle that had been abandoned alongside the freeway and towed it to the storage yard, waiting for its owner to retrieve it. When they didn't retrieve it, we could file a lien against the vehicle for the value of towing and storage fees. This is how I came into my first car. It was a Chevrolet Vega, much like the one I had helped unwrap from the tree the last summer. The Vega was a perfectly awful product of General Motors in the 1970s, meant as an answer to the oil crisis and the high gas prices of that era. It was an attractive car with a sporty stance but a perfectly awful 4-cylinder engine putting out 80 unreliable horsepower. It may have been a 1970s British car for its reliability. But it was a well-known fact among the cognoscenti that they were designed to accept a Chevrolet V-8 engine, and many conversion kits were available.

The doomed car we pulled from the tree was such a conversion, and the engine and conversion hardware were still in the towing yard. Despite my superstitions, I pulled this engine and installed it in my new car. We estimated it had about 500 horsepower, a lot for a small car and a 15-year-old kid. I didn't tell my parents about this until it was too late for them to object. I showed up later that summer in San Jose with my new car - painted and rumbling into the neighborhood like a Gladiator arriving at the Coliseum. It shook the windows in the neighborhood, and there was no doubt among the neighbors about when I was coming and going.

When I arrived home with my new machine, I had my driver's permit and would badger my mother to go out driving with me. My poor mother had become the new unwitting Patron Saint of Petrol and would mostly go on these drives without complaining. On the day I turned 16, I went to the Division of Motor Vehicles and got my driver's license. The instructor that took the ride with me only had one arm, and it was rumored that he had lost this during a student test. We all feared the one-armed instructor, who routinely failed young kids for even the smallest of errors or infractions.

I showed up with the yellow mechanical beast and suspected he did not want to get into it. It was well finished and clean, but the car screamed hot rod with its bright yellow paint and orange and red stripes running along the bottom of the vehicle and around the trunk. And it was loud. The engine had such a big camshaft that it shook the car at idle. To my surprise, the one-armed instructor got into the car, and I made no errors except for using a turn signal in the parking lot. I had a lot of car control stemming from my years of driving go-carts at the limit of tire adhesion, rolling them over, fixing them, and driving them even faster. My brother-in-law had also tossed me the keys to his beautiful 1969 Chevy El Camino at the ripe age of fourteen, and I spent several summers driving it around Yucaipa, running errands, and fetching lunch for my car comrades. Now that I had my driver's license, the world was my oyster.

Driving was a freedom that I had never known before. I would take the car on long drives up into the Diablo Mountains east of San Jose. These unpopulated areas had some of the best places to test the vehicle's handling and sightsee. The hills were a beautiful green in the spring and a mellow golden brown in the late summers. This place became an oasis for me and my car as I dealt with another year of public school and all the oddities that this school afforded me.

The racial tensions at the school were thick, which I was not used to. I came from a racially mixed background, with my paternal grandmother and grandfather having Native American heritage. It was not very obvious to me, despite me being olive-skinned, but my father looked like he belonged south of the border and sometimes encountered problems coming back from brief visits to Tijuana.

San Jose was altogether another situation. At this time, there was forced bussing of students between school districts which was someone's idea of an excellent way to create peace and harmony. It didn't work well, and I found the whole high school experience there terrifying. The low point of this time in school was the 1981 riot between two racial factions, and later that year, a Hispanic student set my hair on fire in class. I learned to be quiet and on the lookout for trouble. In any case, I still had the meanest car in the school, earning me a modicum of respect among all these kids.

One of the people that I befriended during this tumultuous time was Colin. He was one of the few kids I had ever met that was as insanely in love with machines as I was. When I met him, Colin was swapping out the anemic six-cylinder engine from a Datsun 240Z for a Chevy V8. The same bug had bitten him, and we became instant and fast friends. As soon as the daily school grind was dispensed with, we would work on each other's cars. It didn't matter whose car needed the help. We both yearned to be a part of this world again and to experience the sacred bond that comes with other people.

Colin introduced me to drag racing. We had a drag strip near my home, Fremont Raceway, in Fremont, California. I went with Colin my first time to the track; it was one of the most electrifying experiences of my life. We watched top fuel dragsters pounding the pavement with thousands of horsepower and burning nitromethane. It filled me with an excitement that was impossible to describe and impossible to replicate anywhere else. The power of these machines was so intense that you could feel it in your stomach, and at times it would cause your hair to vibrate as the dragsters passed at high speeds. Standing there watching this spectacle, I knew I somehow had to become a part of this. Much like years before, when I wandered into Joe's Machine Shop for scrap metal and adventure, I walked through the pits where the mechanics and drivers congregated. I was amazed that the pits were open for the public to wander through and even touch the machines. It was the mechanics that I liked to talk to and not necessarily the drivers. Some of the drivers were the mechanics, and those were my favorite ones to talk to. I had a million questions. My most important question was how to join this whole racing scene.

Colin mentioned that on Wednesday and Friday evenings, the track opened for what they called 'bracket racing.' This category allowed you to bring whatever car you had and run it on the drag strip. The key was understanding what time it took your car to go from a dead stop, traverse one-quarter of a mile, and then drive consistent with what you declared the time to be. Slow cars would go first, and the fast cars would have to wait while the other cars rocketed away from the light. When the faster car's light turned green, it would rocket away and try to catch the slower car, which would now be halfway down the track. If both cars ran their declared or dial-in time, they would cross the finish line simultaneously.

Most slow street cars in the day took 15-20 seconds. Streetcars that were considered 'fast' took 10-15 seconds. Go below 10 seconds in the quarter mile, and suddenly, complete safety gear, roll cages, and other specialty equipment were required along with a special license. The ground-pounding top fuel dragsters were running under 6 seconds. It was hard to imagine running that fast, but it wasn't hard to guess, bringing my new car to the races and trying it.

My first time on the track was different than I ever imagined. There were many 'administrative' things to attend to, like registration, car inspection, and figuring out my dial-in time. After a few nights of doing this, I became a pro. I got so good at it that I became competitive in the local circuit. As I became better at racing, I frequently broke my car. My mother was happy to lend me her bone-stock original Chevy El Camino to drive while my car was damaged. She approved when I took it to the races but didn't know that I was racing it. I removed any evidence of its track time marked on the windshield, but my stepfather eventually became curious about why the rear tires were wearing thin so quickly. He thought it to be an unusual pattern but never quite figured it out or told me he had. He undoubtedly figured out that I was racing the El Camino and secretly approved, as he had a history of that when he was young. Sometimes as a parent, you must smile and look the other way.

My time in San Jose ended in 1982 when my stepfather announced that his company was closing the local semiconductor plant and moving it to Cork, Ireland, and Logan, Utah. We had to move with the plant, and he had a choice of going to either place. I did not favor Cork as I could not imagine continuing with my petrol-fueled adventures and voted for Utah. I knew very little of the place but consulted my copies of the Encyclopedia Britannica, which was the internet of the 1970s and 1980s.

Upon reading the Utah section, I noticed the picture of a statue in Salt Lake City of the Handcart Pioneers. I mistook the Mormon pioneers for the Amish of Pennsylvania. While I might not have been far from the truth in some respects, I was depressed at the prospect of having to contend with the horse-drawn carriages such as the Amish were known for. It was with great surprise that when I arrived in Utah in August of 1982, everyone drove cars, and everything appeared normal. At least things on the surface seemed normal. Little did I know how much of the role Utah would play in my future and how ingrained I would become with what we outsiders came to know as the 'Mormon Culture'?

To be clear, we were very well accepted into the community and were welcomed with open arms. Neverthless there were many odd things that I could not put my finger on. The first noticeable thing was that the entire neighborhood seemed to know all my family and our history before we even moved in. It all seemed mafia-like. The second big thing was how many churches there were and how they all looked alike. Coming from Southern California, I was not prepared for an encounter with such a closed and religious community. In my way, I tried to make peace with it, but after many decades I eventually just came to accept the ever-presence of the Mormons as part of the scenery in much the same way as the snow existed on the mountains. The snow would come and go, sometimes attractive, oppressive, and pleasant. It was all part of my environment, and I eventually didn't notice it unless I had been gone for a long time. After long absences from Utah, I saw many things when I returned that, taken together, made me crazy for a short period. Like many other things in life, they fade into the background compared to the pace of regular life and the responsibilities that come with it.

I spent my final year of high school in Utah. It was a tedious experience, and I was batted around academically by different school districts who thought I needed to retake certain classes based on the rationale that the other districts were 'academically inferior.' This was bureaucrat-speak for 'Not Invented Here.' My whole goal in high school was to survive it and move on to whatever came next. The only thing I was ever entirely sure I wanted to do in my life was race cars. I found friends in Utah who shared this obsession and found the local drag strip in Salt Lake City. Mormon or not, the love of machines and speed transcended all cultures and religions. I used to tell my car brethren that 'the difference between you and me is the temperature of our caffeine'. Mormons were forbidden by scripture from drinking coffee but found drinking enormous amounts of soda OK. The problem was not the caffeine, as the scriptures discussed, but rather the temperature at which they were consumed.

My last year of high school in Utah passed mercifully fast. Instead of my standard return to Lusby Motors in the summer after graduation, I took a job working as a line mechanic at the local Ford and Nissan dealer. It was a welcoming place full of men and machines and that distinct ethos of having a job to do and knowing how to do it. After twelve long years of submitting myself to the will of intellectuals and others in public education, I was ready to spend my summers on something simple and down to Earth. I settled into a long summer of men and machines in Wilson Motors. Fortunately, I stayed away from most of the girls in Utah and didn't have the additional complications of a female companion to concern me. I was frankly more concerned with spending my time with cars. The other thing I noted living with the Mormons was that Mormonism was a sexually transmitted religion. If you stayed in the safe zone away from members of the opposite sex, you could safely stay away from being converted to the faith.

CHAPTER 4

MARS HAPPENS

Going to college, surprisingly, was a tough choice for me. Nobody in my family had yet to receive formal education beyond public high school. It simply was outside the family tradition. Neither my father nor mother had ever attended the University, as the post-World War II generation could easily make a living in the booming post-war economy. My sister had moved from high school to marriage and raising children, although she did go back to get a BS later in life.

Early in my life, my mother decided I would be the son who broke the family tradition and got a college education. Perhaps it was because of my stepfather's influence being a Silicon Valley executive, or maybe it was her motivation to have her children make a better life for themselves. While I remained intensely curious about the world in general, the thought of spending another four years cooped up inside a classroom was not very appealing to me at seventeen. Because of this, in addition to a general sense of rebelliousness, I formally declared to my mother during the summer of 1983 that I didn't intend to attend college. I made this declaration even though I had taken all the university entrance exams and had done very well. Being accepted was not at all the issue. I had attended several summer programs at Stanford when I lived in San Jose, had an enviable grade point average, and had a resume that would have earned me entrance at any school of my choice. Instead, something inside me resisted having such an important decision made for me. I wanted to arrive at this decision myself.

As I pondered my fate that summer of 1983, I began to fancy spending the rest of my life working on cars. For a 17-year-old kid, it was a good life. I was making good money and enjoying myself. After years of badgering and insisting that my future had a university in its path, my mother finally relented on her insistence that I go to college early that summer. As a result, I enjoyed a relatively carefree period driving the Jaguar XKE roadster and racing my V8-powered Vega at the local drag strip in Salt Lake City. After the races, my friends and I would spend Sundays driving the Jag on the abandoned roads of Northern Utah and making breakneck speeds whenever we could get away with it. On these trips, we discovered the top speeds of these ancient machines and the oddball aerodynamic effects that could happen when approaching 150 miles per hour. We were reckless in our abandonment and impatient with our desire to grow up and live our lives. In many ways, we had progressed very little from those early days riding wooden contraptions down the driveway on Second Street.

The 1963 Jaguar Roadster that propelled us in near death
adventures. Author examining the car. Image Credit Bart Esplin

One particularly memorable time in the Jaguar happened on a Sunday afternoon and would become a symbol of my life. Back in

1983, Logan, Utah, was tiny, and on Sundays, all the good people of the world were either in church or at home, tending to what people tend to on Sundays. For non-Mormons like me, the streets and stores belonged to us on Sundays. My friend Bart and I were headed north out of town and were stopped at the last stop light in the city on Highway 89. Ahead of us was about 5 miles of unbroken four-lane highway with very little traffic. As we waited at the red light, I heard the familiar sound of another Jaguar pulling up next to us. As I looked to my left, it was an identical red Jaguar E-type convertible driven by my friend Don West, the local Snap-On tool dealer.

For those unfamiliar, Snap-On was, at one time, simply the best tools in the world, and trucks full of Snap-On tools plied by skilled salesmen would traverse the small towns and cities of America, assuring that no mechanic would ever be able to save any money and that their toolbox would remain their most significant asset or at least their most considerable expense. The Snap-On truck driving up was like heroin arriving on campus. The younger mechanics would all flock there and slobber over the latest tools and imagine how they would afford them. Don West was the guy who sold heroin to mechanics, and with that filthy lucre, he purchased only the most beautiful cars. His E type was a 1964 model with a slightly larger engine than mine, but it was a perfect car. Don's wife was in the passenger seat and, upon looking over at us, immediately recognized what was about to happen. Don smiled at me, turned straight ahead, and took on serious regard as I told Bart to 'buckle up.'

As the light turned green, there was no burning rubber or the howl of dragsters. One did not do that in a Jaguar. Besides being beneath the behavior expected of such a fine vehicle, you might break the machine if you treated it too harshly. Instead, we accelerated steadily under full power, quickly shifting through the gears and revving the big six-cylinder motors to 7,000 RPM. The combined howl was most impressive, and the cars quickly began to top 100 MPH. Bart assumed the position of holding down the top while Don and I battled for supremacy. Behind us was a gigantic dust and exhaust gas cloud as we headed north on Highway 89 at supersonic speed.

Several cars saw us in the rear-view mirror and dove into the turn lanes to get out of our way, and we blasted through, focused on each other and the road ahead. Neither car could out-accelerate, the other being perfectly restored to factory specifications and thus exactly matched in performance. It was more of a contest of willpower, courage, or stupidity as some might rightly argue, to see who could find the top speed. We approached 140 MPH in a 55 MPH zone and far surpassed the felony speed zone. Somehow that didn't matter as the moment enveloped us, and our lust for speed completely controlled our thinking and actions. Within a short few minutes, we approached the next town, where even we recognized a need to slow down. As both cars descended below 50 MPH, Don raised his hand into the airstream, waved a stylish 'goodbye', and raced on the north. I turned on Center Street, heading to our nice quiet Sunday dinner with Bart's parents. Neither of us mentioned the impromptu race but both of us knew that this was so deeply in our blood that this would not be the last contest of this nature that we would experience.

I had a girlfriend that summer, but she was rather jealous of my cars. It seems I spent more time and money on the cars than on her. Had I been wiser then, I would have ended the relationship right then and there. However, I was a typical seventeen-year-old male and was driven to madness by maintaining a girlfriend. Our relationship that summer became increasingly strained by the cars I owned and my love of them.

Our last time together was quite memorable. I picked my girlfriend up for a date at her home in the Jag. It was a lovely mid-summer evening with temperatures in the 80s. I had the top down, but in a concession to her feminine sensibilities, I rolled up each door window to cut down on the cabin airflow. I was rather proud of my concession. After she got in the car, the complaining commenced. She didn't like the seats. She didn't like the noise. She didn't like the wind in her hair. After about twenty minutes of this, I finally had my fill of complaints as I got closer to town.

I pulled the car off to the side of the road in front of a remote farmhouse. I leaned over, opened the passenger door, and moved back into my seat, hands on the wheel and looking straight ahead.

"Get out!" I said firmly. "You can go into that house and call your dad to pick you up and explain why you don't like this car." She argued with me for a few seconds trying to convince me that she was more important than the car. I kept looking into the distance as if I hadn't heard her. She finally got up and out of the car and slammed the door so hard I feared it would break the glass. Had she been any angrier, I would have expected her to run at the car with her head down to head-butt it like an angry steer. Fortunately, she began walking to the farmhouse, and as she knocked on the door, I drove off.

The thought of going to college began to re-enter my mind as the summer faded in early September, but I was still undecided on what to do. The big push into college by the forces of fate came in a one-two punch. The first push came when my good friend from Wilson Motors, Doug, was killed by a train while testing a car. He had taken the car I was supposed to test drive and did the test for me because he needed to run an errand. While driving, he had not noticed that a train was approaching a road crossing, and he was broad-sided without ever seeing the train. The impact broke his neck, and he died instantly. Seeing Doug die like that suddenly made me think of my mortality and what I truly wanted to do with my life.

The second decisive punch came in an auto part store the following week. Late in the afternoon, I addled up to the parts counter only to see that my high school auto shop teacher Kay Gilgen was also standing there waiting for a part he was buying. We both smiled, said hello, and shook hands. He was a good man and a hard-working one. He taught high school by day and ran a farm by night to make ends meet. Kay had big broad hands that showed evidence of years of scars and hard work. Kay asked me what I planned to do with my life now that I was no longer in high school. When I explained that I planned to be an auto mechanic at Wilson Motors, he looked me straight in the eye and said, "Boy, you are way too smart to be doing this the rest of your life." He continued with a severe look on his face, "This business of working on cars will make an old man out of you fast. You are smart enough to design these cars and not be the guy who must repair them. I don't want to see you in here again telling me that you passed up going to college, or

I might have to kick your ass and kick some sense into you while I am at it". I smiled and laughed at his remarks, but I could tell he was serious. He reminded me as he left the parts store that he would be looking for me and I had better attend the University, or I might be in trouble. I have always considered that conversation one of my life's most influential.

I was insecure about my ability to thrive in the university and didn't think I had what it took to succeed in college. As I returned to work that afternoon, I saw the Ford dealership I worked in with a new set of eyes. No longer did I see the repair bays as places where marvelous machines awaited my attention. Instead, this mechanical hospital seemed more like a dungeon where I might misspend my life if I never had a plan to escape it. I could see exactly what my shop teacher Kay told me, and I could see that even a place full of cars was not big enough for what I wanted to do. I enrolled in the local university the next day.

Utah State University, located in Logan, Utah, was sleepy in the 1980s. Logan is a small town in the mountains of Northern Utah that is achingly beautiful in the summer and stunningly cold in the winter. The mountains rise some 5,000 feet above the high mountain valley floor to reach altitudes of 10,000 feet or more and are covered with snow most of the year. The University is located at the mouth of Logan Canyon, whose main feature is howling winds every morning as the sun rises and pushes warming air down the canyon and out onto the University. Even the trees at the mouth of the canyon told the story of the wind as they all listed to the west at about 30 degrees from vertical like a slightly drunken sailor. By late morning, the tsunami of winds subsides typically, and life begins to emerge from the buildings and enclosures. The student population of USU was around 3,000 people at the time, and its main output seemed to be business, forestry, and engineering students. It was a quiet little campus and seemed to have no drama to speak of and a comforting air of seriousness about it. I liked that.

Given my interest in technology and science, especially my experience designing and building a stereo amplifier, I decided to major in electrical engineering. This seemed like a strange choice, given my love of machines. Still, given my success with building electronics and my stepfather being an electrical engineer, this seemed more logical. It wasn't the right choice as it turned out. Electrical engineering would include a lot of math, physics, and computer programming. I was fair enough at math but had not prepared for college by taking anything advanced in high school besides history and music. I loved history and music in ways I could barely describe and found it easy to excel in those areas. I read history books for amusement and played the piano every spare minute I had. Despite this, I knew I would have to be seriously talented and lucky to make a good living pursuing a career in music or history. Engineering made much more sense and there was a strong demand for engineers fresh out of college.

There was thus a fair amount of competition to get into engineering school, but somehow, I was accepted. The program was split into the 'pre-professional' program and the 'professional' program. My cynical side cast aside such silly labels and thought they meant nothing. However, almost anyone could be accepted into the 'pre-professional' program, which constituted the first two years of the four-year engineering degree. The second part of the degree, the last two years, had competitive entrance criteria, and since there were so many aspiring engineering students and a limited number of slots available. I have never understood the practice of colleges and universities to limit the number of students. I always carried a more blue-sky-capitalist attitude and would have expanded the teaching roles to accommodate the demand rather than creating some 'weeding out' program to limit the student population artificially. As life passed, I discovered that this attitude separated me from 95% of the population. Nonetheless, I faced a second acceptance barrier two years into my education which was required based on grades and extracurricular activities. I honestly didn't know if I was up for it then, but I decided to try it.

My first year of college was a complete disaster. I needed help to adapt to this new life. My fellow students were people to whom I could not relate. The professors seemed to exist in a different dimension than anything I was ever exposed to. The things they thought were funny were hardly even understandable, let alone amusing to me. I felt like a foreigner in a strange land. It took time to develop survival skills. On top of that, I had to take 'remedial' math classes instead of jumping straight into calculus like the rest of the students. I felt at an immediate disadvantage because of this. It got worse. Unlike many other college students, I had not taken any physics or chemistry in high school and was dropped straight down into the middle of something that felt like a foreign language. I have to assume that this lack of preparedness was partly due to my parents' lack of experience in a university and inability to advise me on how to prepare for college properly. I was on my own here to sink or swim. I was paddling hard and taking on water by the end of my first year. I felt like a sinking ship.

The end of my first year could not have come quickly enough. My brother-in-law called late in the spring to ask if I would work for him that summer. This would mean that I would have to quit my job at Wilson Motors, but the lure of extra money was good enough motivation. Early that May, I drove my 1963 Chevy II 900 miles across the desert to Los Angeles, where my job awaited. The car was a rolling restoration I worked on between studying physics and chemistry. I still found working on cars to be something that brought peace of mind and allowed me to think deeply without interruption. My 1963 Chevy II was evidence of that, and I planned to finish the restoration that summer with the extra money I was to earn. By this point in my life, I had learned to rebuild engines, do paint and bodywork, and generally repair anything on the car. And because of some of my electrical engineering studies and practical experience building an amplifier, I could fix any auto electrical problem. That summer, I was consumed with two major jobs that resulted in much income. I became an expert at re-wiring MGB sports cars whose wiring systems were notoriously unreliable. I also managed to repair several cars that had engine fires that were minor in nature and repairable. In both cases, the insurance companies would pay us a flat rate, and I could complete the work in less than half the time allotted. I made more money that summer than the average American made the whole year. At the end of the summer, I returned to Utah with a beautifully restored Chevy II, of which I was quite proud. I placed the car in several car shows and won several awards.

My second year of college went scarcely better than the first. I had made enough money to not work during the school year and tried to apply the extra time to my studies. I still could not 'crack the code' and understand how to do well in college. I was struggling to match my peers in preparation, and while this was something I could slowly repair, it was a constant handicap. A positive side of my second year in college was that I started to make friends with people I could finally relate to. One such friend was Larry Warren. Larry was a considerable man towering about 6' 5", sporting an extensive and full dark beard, and always wore a black shirt and sunglasses. He was a non-Mormon who had been raised in Utah, and this gave him an unusually distinct outlook. This was a distinction of great importance at the time as we considered ourselves a minority much equivalent to a racial minority. The Mormon society revolved around church attendance, and if you did not attend this church, you were a part of the environment and not a citizen. The University was a microcosm of non-Mormons in a Mormon world, and we sought each other out like members of a secret society.

Larry had endured a lifetime of this kind of game playing, and instead of being amused by it as I was, he was bitter. His biting sense of humor about such things, our mutual love of cars and coffee, and our common points of view bonded us together. Larry was also a Mechanical Engineering student and failed to understand why I was studying electrical engineering as I was a 'gearhead.' Larry's influence led me to investigate the possibility of switching majors and saving my sinking 2.2 grade point average. By the second year in college, I was in real danger of failing and returning to that now-feared job as a mechanic at Wilson Motors. At the end of my second year, I changed my major to Mechanical Engineering, and this turned out to be an act that saved both my career and affected everything that happened to me in the future.

I added an extra year to my undergraduate education by changing majors and taking 'remedial' math and chemistry classes. On top of that, I had decided to move out of my parent's house on my own and had to go back to work at Wilson Motors to pay for college and rent. This was a bit of a shock to my routines but did allow me moral and ethical freedom that was very much due. Somehow, the work at Wilsons was not as enjoyable as before, and the honeymoon as a mechanic was over. My high school shop teacher's words still returned to keep me on the straight and narrow and doing well in college. I repaired Fords and Nissans for a few more years at Wilson's, installing air conditioners, rebuilding engines, and plowing the snow in the lot. I enjoyed the solitude of arriving at 0300 to plow snow and have the lot cleaned by the time others arrived for work. It also allowed me to head off to class and have my day's work behind me. It was a challenging but satisfying life, and I rather enjoyed the complete dedication to working towards a solitary goal of graduation. By the end of my third year, my grades had come up, and I was accepted into the final two years of my engineering program. I had barely made it with a 2.6 grade point average, but I didn't care if I had made it by an inch or a mile. I had made it and would continue with a dogged determination that would later be a massive source of strength.

The fourth year was a change for me and instrumental in my future. I had taken a summer job at the Department of Transportation in Ogden, Utah, designing roads. This was my first engineering job, and I obtained it through a car friend of mine, Lynn. He was the Road Department Head of Engineering and shared a love of fast cars with me. He saw that I had a job that summer, and my first professional job came along. I was initially uncomfortable with this new working environment from an office but quickly found that my computer skills and aptitude for math made me a quick favorite in the office. I converted their antiquated card reader computer system to an 'advanced' program that used keyboard terminals to make engineering calculations of survey data and road parameters. This was the first time I could see the otherwise useless math I had learned being useful. I also found myself in high demand because I could draft the roads and do the survey computations. I found that I was well-liked and invited to come back the following year.

UDOT, as it was known, short for the Utah Department Of Transportation, was a place of unusual personalities alongside rather unremarkable people. My boss for the summer was a rather intelligent and intensely smart-ass man named Bruce Swain. Bruce immediately took to me finding a similar smart-ass personality. He enjoyed that I dished his attitude right back to him. For Bruce, it was a sport.

He was also notoriously politically incorrect, even in the mid-1980s. This was in a time before that term was even in our vocabulary. You could say Bruce was a man's man, but like the 1960s, behavior in the modern TV series Mad Men—hard-drinking, heavy-smoking, and always chasing skirts. Bruce's sarcasm hid from the outside world the internal pain that he felt daily from his one and only son committing suicide. I could only imagine the pain it would cause, but Bruce's intense teasing and sarcasm were a way to combat it.

I remember one hilarious episode that happened one morning at work. I had gone to the men's room for the morning biology break and was in one of the bathroom stalls. Bruce followed me in there a few minutes after I went in. I could hear him walk around a bit as if he had forgotten what he came there for. Suddenly, I could see his shadow in front of the stall door. In his booming voice, he shouts, "Cantrell - is that you on the crapper? I know it's you as I can see your shoes." Not thrilled being harassed by the man, I gave him my typical "What the hell do you want, Bruce?" response. Not missing a beat, Bruce replied in his best sailor imitation, "Have you ever titty fucked your wife?". My response was deadpan, "No, Bruce - and if you don't mind, I have other things going on here, and privacy would be helpful." Bruce seemed disappointed by my non-engagement and said quite calmly, "Well if you ever need me to show you how .....". These were the kind of men I had grown up around, and despite the crudeness of their language and choice of topics, there was still a place for them in the world, and Bruce helped me see that. But I knew that after spending a summer with Bruce, I no longer felt a part of that world from which I had emerged, and somehow, the world I was heading into would not be comfortable with me and where I had come from.

A Poster Changes Everything

I returned to the University the following fall, resumed my job as a mechanic working for Wilson's again, and dug hard into my studies. This was about to be a year of big change, and I hardly knew it. Somewhere in the middle of the Fall quarter, I came across a poster on the wall as I walked through the engineering building. I almost walked past it. It was a straightforward poster announcing the addition of a new course in Mechanical Engineering: ME595 Space Systems Engineering. While I could hardly understand what 'systems engineering' meant, I noticed that the course was funded by NASA and was aimed at designing a Mars Rover for NASA. For some odd reason, that struck me as one of the most fascinating ideas that I could conjure up.

I had long been fascinated by space travel, the view of the world that Carl Sagan showed us, humanity's place in the cosmos, and the idea that we could send machines to the surface of Mars. As I read the poster, they were inviting students to apply in person to Dr. Frank Redd, who taught the course. I headed home that night with the kind of excitement that I had not had since I rode my go-carts down the driveway in Yucaipa. I couldn't get the idea of designing a Mars Rover out of my head. I was designing it in my head that night and could scarcely pay attention to the homework I had at hand. I even went across the street from my house to the public library and checked out a few books on Mars. My cursory reading that night indicated that Mars was cold and had a thin atmosphere and a rocky surface. Fortunately, we had some images of the surface and therefore knew something about it. Homework could wait for another day or perhaps another decade.

During an extended class break the following day, I headed over to see Dr. Redd. I walked up the stairs to the third floor of the engineering building and down a long hallway. I noted a door ajar in an office where I hoped to find Dr. Frank Redd. I knocked on the half-open door, and a voice responded, 'Come in.' This wasn't your normal voice, mind you. This voice had a decisiveness to it and commanded you to move forward. This was the voice of a full-bird Air Force Colonel. As I opened the door, I was not disappointed. Frank sat at his desk but immediately stood up to greet me with a firm handshake and a piercing gaze. If I weren't so excited, the man would have intimidated me. He was a smaller framed man, around 5' 8" but had the eyes of an eagle and mannerisms to match.

This was my first in-person encounter with a military officer, and his decisiveness and deliberate purpose were evident. His first words were, "What can I do for you?" Not wasting one single second, I explained to him that I had seen the poster in the hallway, and it had piqued my interest. I also explained that while I was not a senior yet, and these were senior-level classes, I still wanted to be considered for the class. Frank listened intently, looked me straight in the eye, and asked, 'Do you think you can contribute to the class meaningfully?' Such directness was not something that I was used to in this University world. I found my mind moving faster than ever, trying to think of the right things to say and avoiding all the wrong things. Somehow, I managed to convince Frank to accept me into the class, and I offered to start before the end of the quarter if he liked. I was on cloud nine that evening as I quickly finished my homework to make extra time to read the books I had checked out from the library. It seemed that my amplifier days were back, and my mind was enthusiastic again. We were going to build something to go to Mars!

I attended my first ME595 class the following week with my regular class load. At first, I was slightly disappointed as they were discussing a balloon for Mars. I was very interested in the rover but soon realized that the class was so large that they split it into rovers and balloons. As I listened, I began to see that NASA had given the University some money to conduct this course to develop an interest in the space field and the next generation of talent for the agency. To this end, they had given the class the objective of designing a «Mars Mobility System" for unmanned exploration of Mars and the students were to present the design in Washington DC that spring to NASA engineers and managers for their review.

This reality filled my head with panic, joy, and challenge. It was what I experienced modern-day in a race car. We had a challenging task ahead of us, and I imagined we might not survive contact with NASA. The two mobility systems were competing within the class to be the chosen system to present to NASA, and we had two teams working on it. Of course, I chose the rover team and appealed to them to help in any way I could. Most of the team members were more concerned with getting the rover safely from the Earth to the surface rather than very concerned about the rover itself. Such a big job designing a rover was. A lifetime of challenges! I took the task of designing the rover, which I felt very comfortable doing. I also had a drafting board and full drafting set in my home and could easily produce the drawings for the rover.

I worked through the winter designing the rover for our face-off at the end of the quarter while my teammates figured out how to safely place the rover on the part of Mars where the terrain was friendly. At first, I was very concerned that we had to produce drawings detailed enough to build the rover. As time went on, I discovered that a 'system design' was a notional concept of the machine based on sound engineering principles, analysis, and experience. I had many things to figure out, including how it would be propelled, how it would be powered, and how fast it would have to go. All these system-level designs led me back to my basic physics, engineering course, and library.

Soviet Mars Rover climbing a sand dune in Death Valley
CA during a 1992 test. Our rover design for the NASA class
had many similarities. Image Credit Jim Cantrell

To some degree, this was a return to my childhood. I remembered what I was doing to build an amplifier and how I learned from reading books. One of the first things I did was to consult the town library and its books on Mars. There's a lot to learn about Mars: its atmosphere, the kind of surface at hand, and the lessons from prior missions that had gone there. How were things designed for Mars before? This would have been the first rover on Mars, with much new design territory to be covered before the design could be completed. I worked tirelessly on the design and developed full-scale drawings of it. It had six wheels and resembled a modern-day off-road vehicle. I would go home in the evenings from a long day of classes, and my reward for finishing homework was working on the rover design. Admittedly, I snuck in a few extra hours of rover design here and there instead of getting my homework done but, in the end, both topics received adequate treatment.

At the same time as a conventional rover design was underway, another group of students was working on a balloon design for Mars. This balloon would float in the Martian atmosphere, which was very thin and about the equivalent on Earth at 120,000 feet in altitude. Because of this, the balloon had to be very large - about 100,000 cubic feet, roughly the volume of four residential homes stacked on top of each other. Because of the large size, the balloon had to be made of the thinnest materials - 3.5 microns. A typical sheet of paper is generally 100 microns thick - almost 30 times thicker than the Mylar material that we planned on sending 50 million miles from Earth and unfurling it while descending at over 1000 mph through the thin Martian atmosphere. No problem!

By the end of the school year, we were as far as we would get with the design of the Mars rovers. The class had two competing designs: a wheeled rover and a balloon carrying a scientific payload. For reasons of both politics and the uniqueness of the balloon, we chose to present our balloon design to NASA at a conference held for us in Washington, DC. This was a big event where we presented the blueprints to NASA engineers, entertained their criticisms, and watched others go through the same process. For a young Mechanical Engineering student, this was a tremendous opportunity and at the same time, presented a real challenge to us to not look like fools.

Dr. Redd had us spend a lot of time putting together a slide presentation and practicing it in front of several local professionals with experience in the space industry. Luckily, we had a strong space industry in Utah with Morton Thiokol, who made the space shuttle boosters that strapped on the side of the flying contraption. One of the more active and entertaining individuals that came to help us young green budding engineers was Mr. Gil Moore. Gil was an exciting person for many reasons. First, he wore an eye patch on his left eye. The rumor was that he had lost his eye in a V2 rocket experiment in New Mexico after the second world war. We all knew that he was from Alamogordo, New Mexico, and this made perfect sense because this is where they tested the captured V2 rockets.

Dr. Gil Moore left with Dr. Frank Redd right.
Image Credit Utah State University

By 1986, he was already at the age when he could retire, and we imagined all the lore we heard about him to be true. Gil was also a profoundly generous man. He believed so much in students learning to go to space by building space experiments themselves that he single-handedly pushed for and funded the Get Away Special container full of student experiments that flew on every other space shuttle. These trashcan-sized barrels are bolted inside the shuttle bay. After the flight, they were returned to the students, who could see if their valuable experiments survived the harsh environment. One of the GAS flights even deployed a small satellite called NUSat. This satellite, built by USU students, set off the modern small satellite craze and led to today's market of hundreds of small satellites being built. In many ways, it is not an exaggeration that Gil Moore and his GAS can satellite set off the modern 'New Space' industry that is currently the rage. The satellite was built from Radio Shack electronics and put together on weekends by professionals and students for several thousand dollars. This was utterly unorthodox at a time when satellites were the sole domain of governments and commanded prices in the 100's of millions of dollars.

About twenty of us from this design class traveled from Utah State University to Washington, DC. My college friend Ray Levesque was our student leader on this expedition, and Dr. Redd was our parental supervisor. It was an excellent adventure for all of us. We had yet to go to Washington DC, let alone present something to NASA people. To us, the NASA engineers and managers were like gods. They built things like the space shuttle, the Apollo space program, and Viking Landers and put a man on the Moon. I wondered to myself how we simple students, who knew nothing about the design of these rovers, were presenting our ideas to them. Standing up in front of these kinds of people and presenting ideas was incredibly wrenching and nerve-racking. The fear of being ridiculed in public was intense, but our intense preparation partly offset this.

The presentations went very well. We were very proud of our design and the fact that we didn't use viewgraphs which are presentations printed on transparent plastic sheets and shown in viewgraph machines. Instead, we made our presentations on the University CAD/CAM computers and photographed the screen to produce media-quality slides using a projector. Style does matter. There were a lot of constructive criticisms from the NASA panel and a lot of congratulations from our fellow students from other universities. There are a lot of interesting competing ideas from other universities, but ours seemed to be one of the best. Most of us were nervous about appearing in front of many experienced engineers and managers, but they were excellent.

During my presentation, one of the things that I talked about was the orbital trajectory getting to Mars. During the Q&A period, I received a rather elementary question from a rather ragged-looking older man with crazy hair that pointed in all directions, including the sky. I did not know if he had any idea about orbital mechanics and the laws of space travel, so I explained it very simply, thinking that he was a non-technical NASA attendee. After I explained the idea of using LaGrange points in the orbital transfer and what these points were, he replied with a wry smile, "I know how that works as I flew the IEEE-3 spacecraft into that LaGrange point 6 years ago". I was speaking to the world-famous Dr. Robert Farquhar, a virtual god in the business of orbital mechanics and how to maneuver tiny little man-made objects through the solar system. This was one of the first lessons in my life of how extraordinary minds are occasionally carried in the most unusual packages.

At the NASA conference in Washington DC.
Author on the left. Image Credit Frank Redd

After the presentation in the evening, the team headed up the street to a bar not far from the Capitol steps. One featured item on the menu was yard-long beers. Drinking from a three-foot-long glass was not only a challenge but also an easy way to begin feeling the effects of alcohol rather rapidly. We had seen President Ronald Reagan motor up the street in front of the hotel earlier in the day, and this was still the middle of the Cold War. Our conversations soon turned to politics and the never-ending discussion of communism versus capitalism. It was a different time and place in 1987, and this discussion always took on a life-or-death tone as we mere students pondered the world's fate.

Ted was one of my good friends from the class, and he shared my views as a hardened anti-communist. As we poured more beer, the conversation heated and eventually turned to the Soviet occupation of Afghanistan in full force. Ted mentioned that the Soviet Embassy was not far from the bar where we were drinking, and we should walk over there later in the evening and see what our enemy looked like up close and personal. I decided to up the ante and told Ted that a sense of honor would force me to urinate on the front lawn if we went to the embassy. This quickly escalated into a lively discussion and planning.

While our group was mostly enthusiastic about visiting the Soviet Embassy, few were willing to participate. Ted and I finished our beer an hour later and found a cab to take us to the embassy. By this time of the evening, I was full of Heineken Courage and was fully intent on carrying out my pledge. We arrived at the Soviet Embassy sometime around 1 AM. The roads were deserted, and it was a typically quiet summer Washington DC evening. We could see a thunderstorm brewing over Alexandria, but we were undeterred.

As we walked across the damp sidewalks toward the embassy, we could see through the fence surrounding the embassy. We noted that the building looked more like a wealthy home than the Evil Empire's official building, as Reagan called the Soviets. I was surprised not to see a lot of manned security at the gates, which were closed at this time of night. There were some obvious security cameras on the grounds, and you could also see some windows which did not have the drapes drawn. With no lights on inside, I imagined that KGB agents were watching every move we made and were ready to burst through the doors with their AK-47s blazing at anyone who would dare approach the empire's boundaries. The only security we saw was a lone Ford LTD parked down the street with what appeared to be two FBI agents barely awake but observing the entrance to the embassy.

I didn't want to get arrested for public indecency by urinating on the Soviet Embassy. I imagined in my head an outcome that involved a police mugshot of me, expensive bail, and an extended scowl from Dr. Redd, who had brought me to Washington for other purposes. Still, the bravado and alcohol-fueled discussion at the bar about Soviet atrocities in the world required me to save face and carry through on my boast that I would insult the Soviets in the most masculine way I could. I gathered up my courage and approached the fence in front of the embassy.

The Soviet Embassy of the 1980s was a giant granite building surrounded by an ornate metal fence. The large home resembled a baroque architecture more common to central Europe than Washington, DC. I could have easily imagined an ambassador living there or perhaps a captain of industry. The grounds were well-kept with green well-trimmed grass and low shrubs so as not to obscure the view of the sidewalk. As I approached the fence, I looked carefully at the grounds, the backlit sides of the building, and the windows, which reminded me of black eyes on some creature's soul. I kept imagining someone standing at one of those windows and reporting my every movement. Once I reached the wrought iron fence, I looked around with what I thought was a sly scan of the environment, looking for signs of danger, but surely made me look more guilty than I ever imagined. Fortunately for me, the earlier beer consumption began forcing nature to take its course. My urge for physical relief coincided with my disgust at what the Soviet Union represented. The act itself was very easy but lasted longer than I anticipated. I knew, however, that once it had begun, physiology would demand that it be completed in proper order. This made me vulnerable for about 30 seconds.

As I walked away from the fence, I again scanned the environment looking for the inevitable onslaught of Soviet security forces, the FBI, or even the occasional DC police patrol. To my surprise, there was nobody on the streets except Ted. He was laughing hysterically at what he had just witnessed, and this caused me to join him in what amounted to juvenile laughter coming from someone that didn't get caught. As I laughed, I looked over at the Ford LTD, and my undercover FBI agents were still there, but they were laughing, albeit more discreetly. They must have enjoyed the show as much as Ted did, and I waved at them. I got a thumbs up from the agent behind the steering wheel and a broad smile. Ted and I decided to move along on foot before the police might arrive, or the sleepy Soviets came out yelling at the insult that I had just landed on them. As I walked away, I finally began to understand the phrase that my mother frequently used "You look like the cat that just got the canary." Indeed, I had.

CHAPTER 5

BALLOONS ON MARS

An unexpected result of our visit to Washington, DC, was an offer from Frank Redd a few weeks after we had returned. Our team had done very well in the competition, and we were offered two 'visiting summer fellow' positions at the NASA Jet Propulsion Laboratory in Pasadena, California. JPL, as it is known, was a division of Caltech and, since the 1950s, had built most U.S. deep-space probes that had explored Mars, Mercury, Venus, the Moon, Jupiter, Saturn, and Uranus. It's hard to overstate the relevance of JPL to the exploration of deep space and the history of space exploration. This was a holy temple of artifacts that humans first sent to reach beyond the planet and the high priests who spoke about such things in cloaked language, with words like ephemeris, hyperbolic trajectory, and peri-selene. I could not think of a more storied place to work, and I had only read in the library books about the work done there. I doubted that I would ever be chosen for the job, but I nonetheless put my name in for consideration.

Dr. Redd was chosen to send two of his best students to JPL for a 6-month Visiting Summer Fellowship, working hand in hand with real engineers and designers who had flown missions to other worlds. When Frank first picked my friend Ray Levesque as the lead student on the Visiting Fellowship, I was not surprised. Ray was the appointed leader for the class the following year and was a natural choice. The second position went to another fellow student named

John, who was further along in his studies than I was. However, John had a serious girlfriend at the time who objected to him being away for such a long period. In a strange twist of fate, John turned the offer down to spend time with his girlfriend.

I admit to having been perplexed by this decision at the time. In context, however, it was easy to understand, as I would dump a girlfriend for not respecting my car. John was giving up an opportunity of a lifetime to avoid upsetting his girlfriend. As a result, it was an utter surprise to me that Frank picked me instead to be the second student alongside Ray. I had zero expectations that I would ever be picked, and it was one of the highest honors of my life to have been chosen. It was safe to say that Frank had more faith in me than I did at the time, and I was extremely grateful. To be invited to work there was a dream come true that I had never expected ever to be able to fulfill.

Once I was chosen, there was much to do to prepare for what I saw as my first serious job and what I was sure to be a new adventure. Before I left for JPL, located in Pasadena, California, I had a lot of paperwork and phone calls to take care of. My first introduction to JPL was a phone call with my new boss Jim Burke. Jim is a modern-day Renaissance man, if there ever was one. He began his career as a naval aviator at the very tail end of World War II. He and his Caltech buddy Harris 'Bud' Schurmeier had both joined the navy to defend the country, but both missed out on actual combat. They both went to Caltech on the GI Bill afterward and became involved with the budding space program of the late 1940s and early 1950s.

After the Russians put Sputnik into orbit and landed Luna-1 on the surface of the Moon, Jim was put in charge of the nation's first lunar mission named Ranger. Ranger was not designed to land on the Earth's moon but to impact the surface and transmit television images before it smashed into the lunar surface at thousands of miles an hour and broke into millions of pieces. Despite this simplicity, it cannot be overstated how critical this mission was to national prestige and possibly even to national security in the early 1960s. The project was done rapidly.

Ranger first launched in August of 1961 but suffered a launch failure. This was common to the new breed of launch vehicles of the

1960s. Ranger suffered three launch failures before the spacecraft was successfully inserted onto a trans-lunar trajectory. Once a successful trans-lunar trajectory was accomplished, the spacecraft began to fail en route to the Moon. Early speculation was that the forced sterilization of the probe caused reliability problems. Jim later told me that the heat sterilization was eventually replaced by an alcohol rub-down on the pad, followed by a party to drink the remaining alcohol from the launch crew.

As fate would have it, the next three Ranger missions failed before reaching the Moon. Bud Schurmeier eventually replaced Jim as the Program Manager. Ranger was undoubtedly successful due to Jim and his team's early efforts, but Bud got the credit for the success. Jim's attitude in the face of this apparent failure and professional humiliation set an example that served me for the rest of my life. It made me realize that what we were doing in space was hard, that the stakes were high, and that the victorious were not always the ones who deserved the total credit. Jim taught me to smile at the world despite them frowning at me, to believe in myself, and to keep innovating and pushing. In many ways, Jim became one of the most influential figures in my career in just a very short time with him at JPL. He eventually, alongside Bud, who passed in 2009, became my professional mentor and remained so even today.

It's hard to describe the excitement and optimism I was filled with that summer of 1987 as I headed to JPL. I took up residence at Caltech after a long drive across the desert in my 1963 Chevy II that I had restored just a few years earlier during one of my college summers at Lusby's. Everything at Caltech was foreign to me. The attitude was different. The smells were different owing to the intense smell of curry in the shared kitchen at Braun Hall on the Caltech campus. Braun Hall had a bevy of Pakistani students living there, and this was the first time I had met someone from that part of the world. I was unsure how I would live in Pasadena for the coming 6 months, but it all seemed like it would work out somehow. This was college dorm life combined with a new profession that was as exciting as it was new.

Ray and I reported to work on our first day at JPL to get badged and ready for regular entry. One of the more surprising things at JPL that I first noticed was the intense competition for parking spaces. Somehow, the employees needed more parking spaces, and those arriving early were rewarded with spaces near the security gates. Those arriving late had to walk a considerable distance and maybe even be unable to park. It was not unusual to see someone waiting in the car, like a vulture, waiting for the occasional person to leave and leave an open parking space. Because of this, I habitually arrived early and had breakfast in the JPL cafeteria. As it turns out, this was the social hub of the institution, and the grand tradition of all the cognoscenti was to arrive early, have coffee at the cafeteria, and have something to eat. Much project planning, networking, and gossip were exchanged in the morning. I saw several well-known individuals eating breakfast next to me, including my childhood hero Carl Sagan. I used this morning ritual of coffee at the cafeteria to familiarize myself with the day's news and network my way into the organization.

My first office at JPL was sharing Jim Burke's office. He was in Building 180 at JPL, the tall main administration office building. Today, only the most important JPL executives are in that building, which was also primarily true during this time. It was another exciting place to meet people going up the elevator during the day. Jim's office was on the top floor, and the elevator ride was exhilarating. It was a very fast elevator, and by the time it reached the top floor, its deceleration was quite noticeable. Being the young college student I was, I occasionally ignored all decorum alone in the elevator and jumped when the elevator started to slow down. This seemingly allowed me to float in the air and gave me a momentary sense of weightlessness. I did this often enough during my time at JPL to get quite good at it. I almost got caught one day when I fell on return to the elevator floor and was sitting on the floor recovering as I was supposed to exit. The door opened, and several JPL executives were looking right at me on the elevator floor. I made some mumblings and quickly exited instead of trying to explain what shenanigans I was up to for fear of being dismissed from this grand institution.

Jim's office was a ramshackle collection of books, papers, and historical artifacts. I made my office space amongst what I deemed historic documents telling the story of the golden age of U.S. space exploration. This was a time after the space shuttle Challenger exploded, and no U.S. flights were made manned or unmanned. The attraction to reading about the days when these pioneers I was working with built and flew new missions every month was intoxicating and frankly very distracting me from my task. Jim had money from NASA Headquarters to study the possibility of balloons on Mars.

Jim Burke in his younger years. Former Ranger Program Manager and personal mentor. Image Credit NASA

Gentry Lee, who later co-authored several books with Arthur C. Clarke, was our NASA HQ sponsor and occasionally met with us to discuss progress. Jim was focused on the idea of a Martian Montgolfier balloon which would use ambient air, in this case, Martian carbon dioxide, ingested into the balloon as the sun heated it and would create a buoyant contraption that would float above the Martian surface during the day. It would then return to the Martian surface at night as the balloon cooled, allowing a close-up examination of the surface. There were many problems to solve with this technology, and Jim had me working on several of them to find practical solutions to flying balloons on Mars. This kind of problem-solving fired all my neurons simultaneously and kept my mind occupied for most of my waking hours.

The Mars balloon concept. Image Credit CNES

Jim was an avid boater, and one of the first things he did on the Mars Balloon program was to enact a naval experiment of the Montgolfier balloon behind one of his boats. While this was merely a good excuse to get out on the water, it did become very instructive in developing a sense of how these balloons would behave in contact with a surface and how they might heat up under dynamic movement. We also had concerns about how these balloons would behave in descent both during the night and at initial deployment on Mars. It became clear that the Montgolfier design needed a naturally buoyant component enabled by helium or hydrogen to keep the fabric aloft during the night. Eventually, we incorporated that into the design and even experimented with aerodynamic versions of the balloons. JPL engineers looked on as we often would parade large helium balloons suspending ropes below them across the main quad in front of Building 180, and we imagined that they were either envious that we were handling real hardware or wondering what kind of lunatics we might be that security had allowed in that day.

One of the largest problems of this day-night balloon system was the payload itself and its contact with the ground at night. The nighttime winds expected on Mars were sufficient to expect that the ground-contacting payload would be dragged across the surface of Mars. The surface of Mars was not well-known, but it was easy for us to extend our terrestrial experiences of deserts and large physical terrain features to the surface of Mars. When we were imagining balloons on Mars, only images from two locales were imaged by the Viking One and Viking II landers in 1976. These images revealed a mostly flat barren desert strewn with a few boulders. Some suggested that this is an opportunity to anchor the payload at a particular location on Mars and rest for the night, awaiting the morning sun to rise again. However, a large balloon with errant loose fabric pushed by winds can become a very unstable and dangerous force on a small, fragile payload. Our terrestrial experiences with large helium balloons confirmed this. Jim and I realized early on that the balloon's ground-contacting payload had to somehow glide along the surface without becoming snagged by its irregularities, thus avoiding large relative wind forces on the fragile fabric.

It turns out that JPL was one of many groups in the world thinking of a balloon for Mars. The French Space Agency, CNES (Centre Nationals D'Etudes Spatiales) had already begun a project to build and fly such a balloon on Mars aboard a Soviet spacecraft heading there in 1992. The French and Soviet groups had collaborated on an earlier planetary balloon to Venus in the early 1980s, and this system worked well. Two of these balloons were deployed high in the atmosphere of Venus and floated there for several weeks until the payloads no longer functioned. JPL contributed to the mission by tracking the balloons using its unique Deep Space Network (DSN) and helping return the precious data from them. The Soviet group IKI (which in Russian stood for Institute of Cosmic Research) was again working with CNES on a balloon mission to Mars. They even talked to their U.S. colleagues about their plans and invited U.S. collaboration on this mission.

The year 1987, in a historical context, was one where the Soviet Union was undergoing a process of 'openness' known as Glasnost. This meant that the Soviets were willing to open up some of their missions to U.S. involvement. Officially, the U.S. was not prepared to trust the Soviets enough to accept that invitation, but our small universe in Pasadena was soon to test those waters. The Mars Balloon project I found myself in at JPL was soon to become the seed that space cooperation between the Soviets and the U.S. would grow, but I was certainly unaware of this at the time I joined. I was still stuck in the Cold War mindset and saw the Soviets as, if not a threat, not a friendly nation.

As a kid, I imagined the Soviets as described on TV and in movies: nine feet tall and invincible. That was the culture of fear we had all grown up with as young Americans in the 1970s and 1980s. To be sure, I was certain that the Soviets meant us harm as a nation. I soon discovered that, on a personal level, they were far more like we were and certainly far from invincible.

The first Soviet I ever met was Slava Linkin. I thought the last name was a joke when I first heard it, and it certainly provided a sense of irony which I enjoyed. Slava led the Mars efforts at IKI and was a tall and slender gentleman in his mid-50s when I first met him. Oddly, he had a broad smile, smiling eyes, and a beard resembling the American President Abe Lincoln. Slava and his comrade Victor Khirzhanovich had traveled to the U.S. to explore U.S./Soviet cooperation in space at a low level. This had been advocated by the then head of the Soviet Space program Raoul Sagdeev who would later marry the granddaughter of former U.S. President Eisenhower. Victor and Slava were in Pasadena for a few weeks that summer I was at JPL, and there was a meeting with them discussing Mars Balloons hosted on the Caltech campus. I didn't know it then, but The Planetary Society and its founder Lou Friedman had organized their visit and were a force to play a large role in my life and the future of the space industry.

The Caltech campus in the summertime is a sleepy place. It was the perfect place to hold clandestine meetings with Soviet KGB agents, or so my mind imagined at the time. After I received my invitation to this meeting with the Soviets, my mind went wild, imagining what it meant, where it would go, and what might happen as a result. I could easily conjure up disastrous scenarios with me being dragged out of Braun Hall at 0200 by armed FBI agents and interrogated until the wee hours of the morning. I talked to Jim Burke about my fears, and he assured me that this was nothing to be feared and that I might find it interesting. The latter part of his reassurance was a massive understatement.

We met Monday morning in a conference room near Bruce Murray's office at Caltech. Bruce was a Caltech professor but had previously been the JPL Laboratory Director during the 1970s and early 1980s, during whose tenure the twin Voyager missions began their tour of the outer planets. Bruce had a near God-like status in my young impressionable eyes and was hosting the meeting. Other attendees were equally interesting. One individual, Lou Friedman, first appeared in my life at this meeting, and little did I know how much he was to guide me and change the outcomes of much of what I did professionally.

Lou was the Director of The Planetary Society (TPS), a private space advocacy group privately funded by citizens. This was, in fact, an early version of crowdfunding. I maintain that they pioneered much of the modern private space industry with this model. TPS is headquartered in Pasadena and has a close working relationship with JPL. Lou had worked at JPL for Jim Burke in the 1970s as an astrodynamics specialist and was also the father of the modern-day solar sail, having led the NASA solar sail mission study to explore Hailey's Comet. Lou, Bruce, and Carl Sagan together founded The Planetary Society. The mission of TPS was to 'stir up' the more conservative and stodgier NASA and, to some extent, the JPL bureaucracy by provoking these institutions into action where they might otherwise be tempted to remain static. They saw it as public advocacy for space exploration at NASA but created a tremendously important private voice in conducting space missions.

We all filed into the slightly musty-smelling Caltech conference room that Monday morning with high expectations. Almost twenty people were there, and few were familiar faces. The group, I thought to myself, was a rather unusual mixture of people. I had imagined beforehand that my first actual engineering meeting would have mostly engineers dressed in white shirts and ties and sitting neatly around a conference table. Instead, we had something more akin to the famous Star Wars bar scene with an eclectic mixture of people from all walks of life, and some even dressed in a somewhat dramatic fashion. Several older people had a faint appearance of academia, while some would be best described as mountain climbers, complete with climbing boots and shorts. As it turned out, these were the planetary geologists, and the most notable of them was Tom Svitek.

Tom and his family had escaped from the 'workers' paradise' of Czechoslovakia. Lou Friedman sponsored them to become U.S. citizens after spending time in a refugee camp in Austria. Tom would later collaborate with me on topics ranging from SpaceX to LightSail. For the moment, he was a foreign-sounding and appearing individual. Alongside Tom and his mountain climbing attire, a few traditional-looking engineers wore white shirts and ties, as it turned out. I could pick out our Russian guests easily by their accents and manner of dress, unlike anything you could buy in the West. I would not have put it in a higher or lower social status, but you knew it was not bought at the local JC Penney. The colors were somehow grayer than anything we could buy in the west, and the fit was sloppy. I had heard stories of the popularity of blue jeans in the Soviet Union, and after seeing these clothes, I could understand why.

Most other guests were Caltech students and JPL engineers, except for one scientist from CNES, Dr. Jacques Blamont. As it turned out, Jacques had a fascinating life behind that still French visage. He had been at once Science Advisor to the French President and was one of the first to suggest that modern-day Kourou would make an excellent place to launch French rockets. Today, he was one of the largest advocates of international Mars missions. He later explained that he had survived WWII Paris by hiding his Jewish heritage from the occupying Germans. Then once the Americans replaced the Germans, he learned English by reading detective novels left behind by the GIs. He was deeply involved in space exploration since the 1950s, along with several other people in the room, including Tom Heinsheimer.

Tom is an American engineer who had worked on the Atlas launch vehicle that put John Glenn into orbit, once attempted to fly around the world in a hot air balloon with Malcolm Forbes, worked with the Israeli effort to develop the Jericho missile, and later worked with Jacques in war-torn Algeria to fly the French Veronique launch vehicle. He escaped Algeria to get his degree in atmospheric science at the University of Paris and became an eternal partner of Dr. Blamont in future activities like the Mars Balloon. Tom has a sharp wit that cuts like a knife, affectionately wears sunglasses during indoor meetings, and invariably always has his sleeves rolled halfway up his forearms.

At the time of this meeting, I had no idea who I was sitting with and their roles in the space program's golden age. Had I known at the time who all these people were, along with the deep background and extraordinary provenance of most of them, I might have shrunk from the room and disappeared. However, to my young twenty-something mind, these were people that I didn't know, and we were there to discuss what the Soviets and French were planning on doing for Mars exploration.

Bruce Murray stood up and began the meeting. Bruce had a very commanding manner and animatedly spoke with his hands. I was immediately impressed with his simple logic and ability to drive directly and quickly to the point of discussion. He began by discussing the importance of this meeting and how the participants were chosen because of their roles in national Mars exploration programs. He was here to understand how we could all work together independent of national boundaries and old rivalries. This was particularly interesting for me as Bruce once was an officer in the U.S. Air Force and later played pivotal roles in the intense competition between the Soviet Union and the USA in space by leading the first Mariner missions to Mars and the inner planets. Bruce was one of the most charismatic individuals I had ever met up to that point in my life, and I was rapt listening to him.

Slava Linkin stood up after Bruce and began addressing the room in his heavily accented English but with a smile that could never be removed from his face. This smile gave him an air of friendliness I had never expected from a Soviet. His comrade Victor Khirzhanovich accompanied him to the meeting, and Victor also smiled, but perhaps it was with less sincerity than Slava did. His smiles usually accompanied a humorous remark adding to what Slava was saying. I had always imagined that any of these groups traveled with KGB minders, and I had picked Victor out as the likely KGB in the team. Victor's demeanor was quite serious, and his appearance was unusual. Only later did I learn to look for telltale signs of the quality of clothes and gold fillings to ferret out those Russians serving the KGB masters back in Moscow. For now, Victor fit the role, and I watched him carefully as if I was somehow helping to safeguard national security. Several years later, I got to know Victor well, and he began understanding the depth of my non-conformist Behavior. After a meeting, he notably mentioned to me that "in Stalin's days, they came and took people like you out and shot them." For now, Victor was the one I was worried about shooting any one of us.

As Slava talked, he detailed the Soviet missions to Mars they were planning. It was a two-spacecraft flight set to launch in 1992. Each would be launched aboard a gigantic Proton rocket from Baikonur, the Soviet launch site in Kazakhstan, one after another. They would use the spacecraft design that was recently launched for the Phobos mission. One of those made it to Mars, and the other one was accidentally 'turned off' by controllers halfway to Mars. In classic Soviet fashion, both missions were deemed officially 'terminated by meteoroid strikes.' Both. At different times and millions of miles apart. Bruce Murray brought up this rather unlikely explanation to Slava, and his response was, 'This is the official response'. Slava continued with some more commentary, all the while maintaining his very broad smile 'While we know in the Soviet Union officially accidents and errors never happen; this is such the case here'. Bruce was not buying this explanation and chose to back off rather than press his point home.

An interesting characteristic of the solar system is that flights to Mars only practically depart every two years. Thus, the 1992 missions would have to be on schedule lest they wait another two years to 1994 or, worse, 1996. Slava had brought some viewgraphs, which were transparent slides projected on the screen with projector light. They visually depicted the mission launching from Earth, going on an extended cruise of 18 months, and eventually arriving at Mars. When the Mars 92 mission, as they called it, arrived at Mars, it would deploy a rover, two penetrators, and a balloon from each spacecraft. We expected to see two rovers, four penetrators, and two balloons on Mars in just six short years.

Had I not been so taken with the audacity of the Mars 92 mission, I might have pinched myself to check that this was not all a dream sitting in the same room with these people and listening to this conversation. The first time I met Lou Friedman was in this meeting, and he came representing the interests of The Planetary Society. Lou seemed rather ramshackle at the time with his large head of curly hair, thick glasses, and rude Bronx-style personality. In another life, Lou would have made a great car salesman, and in this meeting, he was all about selling the idea of working with the Soviets to us more skeptical US engineers. The benefit, he pointed out, would be to help bridge the gap between the Soviet and American people. While this seemed like a good idea, I was still a child of the Cold War, and this whole scene seemed too surreal to happen, let alone taking Lou seriously. Lou is a Russophile, so much to the point that he would travel there, spending weeks at Slava's dacha and even learning the language. Outside of military men and diplomats of the era, this was unheard of and added to an air of mystery about Lou. Ironically, Lou would become the unspoken father of today's 'New space' despite his un-capitalistic view of the world and only slightly hidden sympathies for Soviet communism.

Jacques Blamont next got up and spoke to the French contribution to Mars 92, which would be the balloon – or Aerostat Martien as he termed it. Jacques spoke in accented but excellent English and exemplified everything I had imagined a proper Frenchman should be. His manner of speaking belied the pride Americans associate with the de Gaulle era French culture. His wit was about as sharp as it could come as he made passing jokes about the stability of the Soviet Union and how this might be our last chance to do something together. Jacques would add in many asides during his explanation of the French balloon discussing trips to Baikonur, incidents in Moscow, and how several Soviets behaved when they came to visit France on official cooperative programs. I was fascinated as Jacques described the Montgolfier balloon and its challenges and instead made it known that the French Mars Balloon would be a single helium envelope with just enough solar heating that it would rise 3-4 km above the surface by day and descend slowly and majestically to the surface as the Martian sunset arrived.

At the time, the French Space Agency CNES (standing for Centre Nationale D'Etudes Spatiales) was to build the balloons and the gondola, and IKI would build a ground-penetrating radar mounted on the ground-contacting part of the balloon flight train. The belief at the time was that Mars had a hidden cache of permafrost, which contained much of the ancient water that had once covered its surface. With its ground-penetrating radar, this balloon mission would prove that the water permafrost exists and would map it globally via balloon. The Soviet design for the ground-contacting payload looked more like a piece of Samsonite luggage than a scientific instrument. Jacques made several off-color remarks about it getting lost in the Aeroflot baggage handling on the way to being integrated at the launch site. Jim and I were more concerned with how such a piece of luggage would slide along the surface and how it would traverse cracks and crevices, which were undoubtedly present. When Jim asked his questions, Slava's response was much like the attitude toward accidents in the Soviet Union: 'According to Soviet science, there are no such crevices in the Martian surface.' By the smile and laughter following this remark, I could tell that nobody, including Slava or Victor, believed this stupid statement.

At about this time, Tom Heinsheimer stood up and took control of the conversation from his old mentor Jacques. Tom had a long history of piloting manned balloons and held a balloon pilot's license. He once piloted a helium gas balloon from the shores of El Segundo, CA, out over the LA Harbor, mapping pollutants over the ocean in the late 1970s. He drew an image on the board of something Jim and I were now familiar with a rope dragged below the balloon as we had done behind his boat. This time Tom drew it operating on the surface of Mars under the balloon. Tom termed it a guide rope and began describing the history of the device going back to European balloon expeditions of the Arctic and how the smooth dynamics of the rope combined with an ability to use several ropes to steer the balloon. He suggested we use the same technique on Mars and begin by figuring out how to build it. Jim and I looked at each other in one of those knowing moments and smiled. The others started arguing. They argued about the mechanical complexity of such a device, how we need to conserve mass on this very lightweight flight train, and how we could never model the dynamics of such a thing. Tom again suggested a balloon experiment with guide ropes versus Samsonite luggage cases to settle the argument. He further indicated that we could pull this experiment together quickly and pull it off by mid-summer.

Lou and Bruce seized upon Tom's suggestion, and the wheels were set into motion. At JPL, we agreed to provide helium balloons and some prototype payloads. Caltech would provide vehicles, and TPS would work the plan and provide volunteers. Lou was incredibly proud of their involvement as it used member donations and highly leveraged that small amount of money to buy things neither JPL nor Caltech could, thus bringing TPS in as a full participant in this new Mars Balloon mission. Tom agreed to bring a hot air balloon, its crew, and himself as the pilot. We were ready by early July and headed out to the Mojave Desert, where we had located what we thought was a perfect place to test: Lavic Lake. This spot had both the necessary terrain and was close to a major thoroughfare being just off Interstate 40 near the Arizona / California border. Jim and I had scouted the place by air, having rented a Cessna 172 in Barstow and flown over the area. It was perfect as it had a sizeable sandy playa, much like the images we saw on Mars: some sandy dunes and large outcroppings of lava. These lava formations extended into the dry lake-like fingers and would make an ideal terrain to test the payloads absent of vegetation that is not present on Mars as far as we know. The only problem with this site is that it was on the Twentynine Palms Marine base. Not being one to give up easily, Lou managed to talk the Marines into allowing us onto an active bombing range if we were escorted by Marine personnel. The Marine Base Commander was sufficiently intrigued by the whole idea and content with the JPL/NASA connection that he allowed us to perform our first Mars Balloon experiments there.

A large contingent of no less than twenty people set out to Barstow from Pasadena late in the afternoon, and all stayed the night at a low-budget hotel in Barstow chosen by Lou. We all agreed to leave the hotel at 0300 the following day and gather at the test site by 0400. The idea was to set up for the hot air balloon flight as the sun rose and to drag several payload prototypes from the balloon as it traversed the lava fingers. The morning air is the calmest of the day, and the sunrise would provide a consistently predictable slight breeze in the area away from the sun as the atmospheric air heated and expanded at dawn. As we gathered at the hotel, we watched the storm clouds gathering to the east of us near the test site turn into an enormous summer thunderstorm, complete with the wrath of God lightning and dramatic winds. Somehow, when the rain made it to Barstow, the tests in the morning using fragile balloons seemed like a terrible idea. Nonetheless, we still planned to leave early in the morning and hope for the best.

Jim and I headed out to the test site in the early morning dark, driving in his 1963 Chevy II wagon. It was a sister to the car I restored and drove that summer but had a lot of room in the back to haul the guide rope that he and I had made by stuffing rope into a large piece of PVC and doing it in such a way that it could be curled up in the back of the station wagon. He and I drove silently together for an hour to the test site, saying very little, but we had much to think about. Jim wore his trademark white tee shirt and jeans and always seemed to have this optimistic smile on his face. Once we arrived at the rendezvous site, we were met by a Marine Corpsman who gave us a safety briefing. He mentioned that this was a live bombing range and that we might encounter live ordinance. He instructed us not to disturb the ordinance and report it to him. He also mentioned several large bomb craters on the dry lakebed floor and that we should be careful of those as they were not visible until right upon them.

Somehow on the way out to the balloon launch site, Jim and I separated from the line of cars. Jim stopped to consult a map with a flashlight and a compass and decided on a course correction. As we were heading in a northerly direction at a relatively high rate of speed, we suddenly encountered one of the bomb craters that the Corpsman warned us about. Jim's station wagon flew momentarily through the air as we dove into the hole, and I could hear all the equipment in the back of the car hitting the roof. I was waiting for it to hit us in the back of the head when suddenly, the vehicle landed. I could hear the engine roar as Jim pushed the throttle to the floor and yelled, 'Hold on!'. All we could see in front of us was a wall of dirt, and then, suddenly, the car rotated its nose vertically, rose into the air again, and landed in a repeat of the prior maneuver. Jim continued driving without even missing a beat and looked over at me, saying, 'That must be the bomb crater that he was talking about.' I was too shaken and full of adrenaline to answer with anything other than, 'Have you done that before?'. Jim responded that he had not but knew that we would be stuck in the crater if we didn't keep going. OK, I learned something new that morning: how to get out of a bomb crater.

As we pulled up to the assembled masses, Jim and I exited the car. I inspected it for damage and only noticed that the two remaining hubcaps were missing. Aside from the dirt, this old car did well to navigate the bomb crater. Neither mentioned it and readied the guide ropes and the Samsonite suitcase payload for the balloon's flight. The hot air balloon crew had already taken the balloon from its container and laid it carefully on the desert floor with the gondola placed to the east. Lou and his volunteers were busy taking pictures, videos and providing coffee and donuts. As I stood back and watched the coordinated action and chaos, I felt deeply connected with the effort going on and noted that this somehow felt like something that reminded me of the spirit of racing cars that I so loved. I somehow knew that this scene could very well be repeated in my life again and again in the future and that I could not be more content with it.

At the expected time, the hot air balloon inflation began. They used a large fan to inflate the balloon envelope, much like we inflated the Montgolfier during our experiments with a forward scoop. This gave the balloon a predictable shape and allowed the fearsome burners on the gondola to begin heating the air inside the envelope. Slowly as the sun rose on the horizon and brought the brilliant light of early morning, the hot air balloon lifted off with a crew of three. We followed alongside the balloon with vehicles and cameras as best we could as they first deployed the Samsonite luggage payload. We could see that its path along the sand was jerky, but once it reached the lava fingers, it became violent in its motion. This caused severe oscillations up the connecting rope to the gondola, and the crew cut it loose in short order. Next, the guide rope was released from the gondola, and as it touched down on the surface, its motion was smooth and predictable on the sand. Once it reached the more rugged lava, it maintained a very smooth motion having stiff bending resistance and integrating out all the rugged surface features of the lava. It was undeniable from this straightforward demonstration that this was the right design for the balloon ground-contacting payload. We followed the balloon for another 2 miles until it reached the far end of the lava outcropping. There the balloon landed and maintained its position to offer rides over the lakebed. This was my first time in a hot air balloon, and it was a magical experience. Once we completed the rides, we declared victory by 0800 in the morning. We packed all our equipment up and returned to Pasadena to regroup the following Monday and discuss the results.

On the way home, Jim and I discussed the guide rope and the very positive results from this test. A long-distributed mass like a rope was essential to the dynamics of a surface-contacting payload. The practical reality of using PVC and rope was one thing for a low-cost terrestrial experiment. We knew that the plastic surface would rapidly abrade on sand or rocks, and the frigid temperatures of the Martian surface would require compatible materials that did not become brittle at -200F. Jim and I agreed to try to find a practical solution to this problem. The next day in his office, Jim was busy taping together some Dixie cups overlapping and connected by a string. As I entered, I stood in the doorway and watched him duct tape the last cup to the string he borrowed from the numerous kites he liked flying on the JPL quad. He looked up at me and, with that broad smile, said, 'This is how we need to build the guide rope'—puzzled, I watched as Jim dragged the 3-foot-long trail of Dixie cups held together by tape and string over the various books, awards, models, and other office debris. The movement reminded me of a snake. If dragged in the right direction, the cups easily rode over sharp edges, and the flexibility of the contraption resulted in a smooth ride over large terrain undulations. My mind began to race as I sought to imagine how we might practically make this device. Jim explained that he thought of this as he watched the PVC pipe travel over the lava flows at Lavic Lake a few days earlier and thought it might be the solution to the ground-contacting payload. The practical design, he said, was mine to solve. In classic Jim Burke fashion, his brilliant mind had conceived of a solution to a vexing problem, and he had turned the 'details' over to me to complete.

The following week was the Fourth of July holiday, and I had decided to head back out to Hemet, California, to see my cousin Mike and his wife Judy for the holiday. Mike came into my life in

1978 when everything else seemed to be going to hell. Mike had suffered numerous life hardships and learned to laugh back at them. Mike taught me to laugh at life, smile at people who had done me wrong, and see the perverse humor in our world. Just being with him and being in his home brought me a sense of peace and comfort. This weekend, however, I got a homework assignment with me.

In the week after Jim and I witnessed the great Dixie Cup experiment, I began to think about the design of the actual guide rope and how it might be implemented for something on Mars. The temperatures on the warmest day were well below the freezing point of water, and the night times would plunge to well below -100C or (-150F) which would terminate most plastics or flexible materials. The design would have to be made of metallic substances to withstand the temperatures and the abuse of dragging along a rocky surface at speeds up to 35 miles per hour.

After hours, I went to the local hardware store in Pasadena and bought twenty rectangular pieces of stainless steel and about 20 feet of chain. This was going to be my prototype Mars payload. In a manner utterly reminiscent of my go-cart days, I laid all of this out on Mike's garage floor and began to roll each sheet into a tapered cone with a 6-inch opening on the bottom and a 5-inch opening on the top and riveted along the seam. Mike watched me in bewilderment and commented that NASA might now need to designate his garage someday as a place where Mars probes could be made. Mike was an accountant and understood little of what I was up to. He knew me well enough to be aware that since I was a child, I would spend endless hours building things, and somehow, someway, they would turn into something useful.

I chained all the segments together, overlapping, as Jim had done with the Dixie cups. In less than three hours, I had my first Mars guide rope. I tested it out in Mike's front yard, and his neighbors must have wondered about my sanity as I drug a shiny fifteen-foot-long contraption over bushes and fences in front of the house. When I was done, I could collapse all the cones down to a single unit less than four feet long and place them in the trunk of my old Chevy II. On Tuesday, I headed back to JPL with my prototype.

We hung the guide rope from a small tetrahedral helium balloon inside a small warehouse at JPL. The overall aesthetic of the device was pleasing. One of the things that Jim taught me was that with aircraft, if 'it looks good, it flies good, and if it looks bad, it flies bad.' Having little experience among us with ground-contacting balloons on Mars, we, of course, would have to test this one to decide whether it would work well.

Later that day, we took the balloon on a walk with a leash and dragged it across the JPL campus. Jim looked like he was taking his pet balloon for a walk. It efficiently behaved well on its first maiden voyage crossing curbs and the occasional traffic barrier. It glided smoothly over pavement, grass, and concrete. We ended up in the main JPL quad and allowed the balloon and guide rope to rest while we ate lunch. The gentle breezes moved the contraption through the quad, occasionally visiting other lunchtime occupants but never getting too bothersome. This was a success, but the time had come to take it back to Lavic Lake and see how it did on that Mars-like terrain.

Somehow the combination of August and the Mojave Desert enticed many of us to pack up Chevrolet Suburban full of balloons and equipment and head out to do testing. This August in 1988 was no different. This time, Lou Friedman had rallied support from CNES to attend our balloon tests, and they agreed to send some small prototype balloons for us to use in our tests. Lou, in turn, had decided to bring a group of Planetary Society volunteers out with us in August to support and film the tests. The usual suspects, including Jim Burke and myself, headed to Lavic Lake in late August. We initially planned to return to the original test site where we had done hot air ballooning. Still, the Twentynine Palms Marine Base had a new base commander who viewed collaboration with Soviets, French, and wild-eyed American scientists with less humor than the prior base commander.

We improvised by operating in the same area but adjacent to the base boundaries. We located roads to access the site, a nice sandy spot to set up the operation, and a path where the prevailing winds could take the balloon and guide rope on its maiden voyage free with the winds. I somehow led the recovery team to a set of Caltech

Suburbans equipped to go off-road. These were the Geology Dept vans that Bruce Murray had cleverly appropriated for the week by declaring the tests as a 'geology expedition' in the Mojave. I learned a lot of tricks like this from Bruce and admired his ability to work around man-made obstacles like this.

The morning of the test could not have been more uneventful. As usual, the winds were mild at dawn, and everything was in place while we inflated the balloon. The recovery crew and I were stationed several miles downwind and connected with mobile radios. Lou had a crack imaging team who would photograph the test and videotape it. The French contribution was the first balloon to be inflated, a 20-foot-long cylindrical balloon. It was made of very thin Mylar and was fragile to handle. True to form, a gust of wind blew the balloon down during inflation, and its fabric touched the rough lava-strewn surface. The French experiment did not survive. Fortunately, Jim and I had packed some tetrahedral balloons away from our JPL tests, which were put into use. Once the balloon and guide rope assembly were inflated and set free, it moved slowly and surely over the terrain. We received good reports over the radio that it was moving our way as planned and moving quicker than expected. We watched as the balloon bobbed up and down and got larger and larger as it approached us. We could see the mechanical guide rope I had built in Mike's garage effortlessly scale meter-class lava fields and move along as if it were alive. The balloon did not shake or jerk as we had seen with the Soviet Samsonite experiment.

The Mars SNAKE in action on a lava field.
Image Credit Jim Cantrell

We had planned to get into the Suburbans and abuse the machinery in pursuit of our device as Interstate 40 was a mere 1.2 miles away. Having this loose on the freeway was something other than what we wanted to see on the news. Instead, the balloon came directly to us while we were parked on the road, and it was snatched by one of the Caltech Geology students sent with me to ensure I didn't abuse the machines. We all declared success from these tests, and Lou was most happy that we had photographic and video evidence that we had successfully tested a Mars Balloon prototype. You had an interview with LA radio station KFI AM later that afternoon, where he declared that the day of citizen-led missions to Mars was born. As I watched Lou talk over a car phone, then a massive novelty, I realized that this was more than an exciting summer in my life..

CHAPTER 6

✦━━━━━━━━━━━✦

BALLOONS AND
BAGUETTES

R ay and I headed back to Utah State later that year, having
had great success with our efforts at Mars Balloons. Ray had
concentrated on trying to model the thermal aspects of the
balloon while I was building what amounted to the Mars mission
equivalent of plywood go-carts to test the dynamics of a ground-
contacting balloon. The Planetary Society was so excited about this
result that Lou and Bruce spoke with CNES about becoming a 'US
Partner' in this Mars 92 mission with our new guide rope technology.
While JPL had developed it, Lou even offered to continue funding
its development at Utah State after I returned to graduate school that
next year. Frank Redd also agreed to find some department money,
and the TPS Mars Balloon program was born.

Frank was becoming a well-known and significant figure in
space activities at Utah State, already well-known for its space lab,
having launched more than 400 payloads over 40 years. Frank had
been drawn to Utah State by this reputation and the then-budding
small satellite industry, which had its birth at Utah State. NUSat,
standing for Northern Utah Satellite, was designed and built with
volunteer dollars, integrated at Utah State, and flown out of the space
shuttle Get Away Special container (GAS can) in the mid-1980s.
The GAS can was the brainchild of another well-known local space

personality Gil Moore who worked for Morton Thiokol building space shuttle boosters but, more importantly, spent his own time and money to get the GAS cans put in place and flown on the space shuttle. Gil was dedicated to energizing young people into loving and understanding space travel. He has inspired thousands of us in the industry to do good things and give back through the example that he has provided us.

Gil, originally hailing from Las Cruces, New Mexico, but relocated to the small town of Logan, was a big friend of the University and worked closely with Frank Redd. Together they started the AIAA Small Satellite Conference in Logan when few thought small satellites would ever become an industry, let alone the primary driving force in international space commerce. Their first conference had fifty attendees, and I presented two papers there. Gil was an ever-smiling polite character and carries a manner from days ago where politeness and courtesy are valued much more highly than one's intelligence or possessions. He also wore a signature eye patch over his left eye, giving him a distinction worth remembering.

The rumors were that Gil lost his left eye during an early V2 rocket experiment working with the Operation Paperclip Nazi scientists that came to the US in the 40s. I was brave enough to ask Gil if this story were true. He laughed in his inimitable style and said that he wished it to be accurate, but instead, it came from a screen door that hit him in the eye as a child. So much for rumors! Gil later co-founded Globesat in Logan, Utah, with Rex McGill to build a new generation of small satellites. They were, unfortunately, 30 years before their time. Still, their efforts sparked the small satellite Revolution and eventually led to a collaboration with Dr. Bob Twiggs and the creation of the Cube Sat.

I was scheduled to complete my coursework for my undergraduate degree by early 1988 and was considering going to graduate school. I spoke with Frank Redd, and he encouraged me to do so. He also offered me a job at the Space Dynamics Lab during the school year to continue working on the Mars Balloon work I had started. He, too, was intrigued with the whole 'citizen-funded' mission to Mars, and Lou had taken a trip to meet Frank that winter in Logan. The two of them hit it off, beginning a long and fruitful relationship between the two institutions. My only problem was that my grade point average, 2.9, was insufficient to be admitted to a master's program, which required a 3.0. In the fashion that Bruce Murray had taught me, I was undeterred by this and developed a relationship with the Department Head's secretary Joan. I knew that Joan was the secret gatekeeper to the graduate program, and when the decision came, her weighing in on my side would tilt the vote in my favor. I was right, and my relationship-building paid off.

I spent the next eighteen months refining the design of the guide rope and developing ever more sophisticated prototypes for testing. During the winter months, I spent time researching the biology of snakes. I hoped to find mechanical inspiration from mother nature by understanding the physical construction of a snake. I got the idea from watching the videos of our contraption slithering over the terrain in a manner reminiscent of an actual snake. What I found that I eventually emulated in what became known as the Mars SNAKE was an exoskeleton on the outside (like the hard underbelly scales) for sliding easily on rough surfaces with an endoskeleton on the inside (like the snake's spinal column) for longitudinal strength and to avoid structural compromise by a damaged exoskeleton.

We started working more closely with the CNES and IKI in the Soviet Union on the requirements of the SNAKE and ended up with a slender 30-foot-long device that weighed a mere 10 lbs. We iterated the design with stainless and titanium exterior surfaces and quickly settled on titanium for its lightweight, high-impact toughness and abrasion resistance. The only problem was that this was expensive, and a very budget-conscious project being funded by TPS donations and USU Mechanical Engineering research money.

I soon collected a group of students, and we built a flight-like model for testing. We hung it from the University buildings to test for landings, drug it behind a truck and trailer in parking lots, curled it up, and deployed it like it would need to be during stowage on its way to Mars. By August 1989, we were ready for another device test in the Mojave Desert. This time the French contingent from CNES was present, as were the Soviets from IKI and a new group named Babakin. As luck would have it, the International Astronautics Federation (IAF) conference was being held in Pasadena that year, and all of the key players from CNES and Babakin were in town. We took advantage of this presence to meet with them and figure out how we could become part of this mission to Mars.

Babakin was a particular group from NPO Lavotchkin who, during the Soviet times, had built fighter aircraft during WWII to fight off the invading Germans who made it to the factory doorsteps in Khimki, Russia, and had later built all of the planetary probes for the Soviet Union. This was the Soviet version of JPL. Because of Soviet secrecy and Lavotchkin building military satellites alongside the deep space probes, Babakin was the group dedicated to working with Western agencies and individuals.

A man named Raold Kremnev was the leader of NPO Lavotchkin and Babakin. He attended our pre-test meeting in the same conference room at Caltech where we met years before. Both Victor and Slava were there as well, along with Dr. Blamont and several French colleagues, including the Program Manager of the Mars Balloon project in Toulouse, France. Gil Moore also attended this meeting in place of Frank Redd, who, being the eternal cold warrior, could not bring himself to sit in the same room with his old Soviet enemy. People underestimate the power these old hostilities can hold over a person; it was still very alive with Frank.

Roald was a hard-looking man and Soviet-looking in every way I had ever imagined a Soviet Man to look. He had bright blue and piercing eyes that belied little humanity but plenty of intelligence. His hair was cut short and perfectly formed backward as if he was posing for one of the Soviet propaganda posters depicting Soviet workers in an eternal victory. He was also a massive man with a large frame, almost the equivalent of mine. When he shook hands, you could feel the strength in his grip and the sheer size of his hands, which dwarfed my own abnormally large hands.

Raold was clearly in charge, and after Bruce's introductions, it was my turn to present the status of the Mars Balloon efforts, particularly the guide rope design. CNES had already baselined a similar plan for the mission, and we were informally cooperating with the French and Soviets. Lou was the glue that kept all this going and continually provided high-value 'wedge money' raised from mass mailings to TPS members to do things that the rest of NASA could not. This was a TPS show for IKI, Babakin, and CNES. Lou wanted a seat at the table for the Mars 92 mission.

I began the presentation by discussing the Mars Balloon concept's evolution and the ground-contacting payload problem. I showed some images of our hot air balloon flight at Lavic Lake with the Marines. Tom Heinsheimer was in the audience and spoke for a few minutes about the value of testing these balloon systems rather than merely studying them. I then discussed the concept of the mechanical guide rope that we now call the SNAKE. I discussed the snake and its behavior as it dragged over the surface. I then showed Kremnev a video from our test that summer.

At this point, the decision was made to incorporate our SNAKE into the Mars 92 mission. Kremnev began to ask me a series of questions about how it was constructed and how it could be used as a radar antenna to penetrate the sub-surface. He also talked about how pleased he was to see 'peasant wisdom' used to develop such a device. He described that every Soviet peasant knew that driving on rough roads was better above a certain speed as the car began to skate over the rough bumps in the street. Kremnev truly admired our work. I was unsure how to feel about being complimented for my 'peasant wisdom,' but it gave Lou, Tom, and Bruce a lot of fodder for future jokes at my expense.

Lou was the master of getting people together and creating something much larger than the sum of the people he gathered. Later that night, Lou had us all gather at his home outside of Pasadena, with the typical fare of Armenian food from Crown Burger in Pasadena and beer. It was the beer that drew the Russians, I was sure. Kremnev, Blamont, Heinsheimer, and several other CNES personalities were there that evening. Slava and Victor also attended, and there were several conversations. Tom took the lead in developing a plan to fly hot air balloons in the Soviet Union and asked me to join in an expedition there later in the year. We discussed flying balloons in Irkutsk as Kremnev knew of some Mars-like terrain in that region, and he had spent a fair amount of time there. Little did I know at the time that Irkutsk was in Siberia. We walked away from that meeting clearly with a plan to work with the Soviets and CNES, and the cooperation would be through The Planetary Society. At this point, TPS became the center point for American involvement in the Mars 92 mission. As for me, I could scarcely believe any of this was happening.

The next evening, the CNES people invited me for dinner at a seafood restaurant on Colorado Blvd. They wanted to discuss my coming to work for them in Toulouse. Jacques Blamont was championing this idea and was seemingly popular with the project team led by Christian Tarrieu. Christian was a classic Program Manager, pragmatic, and very direct in his speech. I liked him immediately. My family was unsure about my moving to France, and I was uncertain about the whole idea. Ever since I stepped into Frank Redd's office several years earlier, I had felt like I was on a train headed to an unknown destination. Most of the time, I had faith that the destination was an excellent place to go, but sometimes my faith wavered. I gradually exercised more deliberation in reaching essential decisions before me, which was one of those times.

The discussion at dinner began with the usual chit-chat about the differences between French cuisine and American fare. The French have a very high, and deserved, opinion of their national talent: cuisine. Naturally, they like to discuss this pride, especially while dining at American restaurants. I sometimes imagined that the French wanted to visit America specifically to refuel their arsenal of commentary on American life and its direct comparison to the French way of life. Nonetheless, my future French colleagues were soft on me for this culinary event but did pay extra attention to my choice of food. After my salmon arrived, I began eating but noticed that the entire French team was watching me eat. I stopped, looked up at them, paused for a second, and finally asked, 'What is it?'. Christian smiled and, in his usual direct manner, said, 'Well, we are not used to seeing an American who uses the fork in the left hand and the knife in the right hand in a proper European fashion and does not put catsup on the food.' I laughed when I heard this but could see from their lack of smiles that they were serious about the question. I explained that this was a practical way to eat, that I could consume food more efficiently, and that I never liked Catsup. One of the older team members looked at Christian and said, 'That's the funniest thing about Americans. We know exactly who they are, but the real ones always run and hide when visiting them. Little did I know that this would be a regular theme for the next three years of my life.

The post-dinner discussions were about business, but the discussion was all about personal matters. I would later find out that this is the European style of business. It is considered rude in European social circles to dive immediately into business without proper social discussions beforehand. This tends to ruffle us Americans as we are the exact opposite and read this behavior as unnecessarily slow and avoiding the point. Once we had consumed an adequate amount of wine and they finished interrogating me about my personal life, we were ready to discuss the matter of my working for CNES. Christian made it clear that they would like me to come to France when I graduated late that year and lead the development of the Mars SNAKE. They would pay me what I needed, and I could continue the work I had begun in the U.S. Back then, we didn't concern ourselves with things like the International Trafficking in Arms Regulations, also known as ITAR. These esoteric regulations seemed to apply more to the arms manufacturers and not anything with which I was involved. Little did I know that the definition of 'defense articles' under this regulatory authority included almost anything that went into space. While I was not technically violating ITAR by working for the French on a space program, it certainly skirted the line and, in certain opinions, broke the rules. I recall a conversation ten years after this time where an official of the U.S. government accused me of 'exporting my thoughts' that are regulated by ITAR. I found this amusing, and it helped to guide my disdain for government regulations ever since.

That night we devised a deal in principle for me to work at CNES in about six months. I made a trip later that fall to France for the first time to visit, and we made some plans for my eventual arrival. I was going to be a Project Manager at a very young age and working in a foreign space agency. This is certainly something that I had yet to foresee back when I pondered whether to go to the University. Looking back on the events of this summer, it was a significant inflection point for me personally and for many things that would later come to pass in the industry. It was the birth of citizen-led space efforts pioneered by The Planetary Society, the beginnings of the U.S./Russian space cooperation that would lead to the International Space Station, and one of the seeds that led to SpaceX. It was also a significant shift in my life, but at the time, all of this seemed very normal and not the least out of place. I returned to Logan that fall about the time the leaves were turning brilliant colors of yellow, red, and orange and the time that the familiar drone of university classes began. This was to be my last college experience, having received my bachelor's degree in mechanical engineering and only needing to finish my thesis to obtain my master's degree.

I finished my thesis that winter at USU. It was a quiet but busy part of my life at that point. My first daughter had been born earlier that year, and things were settling into a fierce but predictable pace. After the new year, I began preparing for our new life in France. There were a lot of things to work out, like contracts, housing, vehicles, and other minor details of life. As it turned out, The Planetary Society would sign an agreement with the CNES for my services and pay me. They supplemented my paid work with donations from members who believed in the mission. Lou, the consummate car salesman, spun this tale into a once-in-a-lifetime opportunity for citizens to participate in a mission to Mars privately and know that their money went directly into making it happen. TPS would send out mass mailings several times a year quoting Lou talking about the importance of international cooperation in space and how even small donations would allow The Planetary Society to contribute meaningfully to the mission. For Lou, every 25-dollar check would matter, and TPS did a great job of stretching every dollar that went into the effort with volunteers and Lou's inherent thriftiness.

Lou hired Bud Schurmeier as a consultant to work with CNES on this project bringing his decades of experience running the Ranger, Mariner, and Voyager missions at JPL. Bud became a vital mentor to me, and I was honored to have his attention and help. For a man who had such a storied career building spacecraft that went to nearly every planet and beyond our solar system, he was an incredibly humble man. He was always serious but with a smile. His simple manners and simple logic reassured me and the choices I would make during my career. Bud came from very humble beginnings and wore no airs of importance despite his highly impressive career. He loved to fly, having been a naval aviator in WWII along with Jim Burke, and spent much of his time chasing thermals and excellent atmospheric conditions in gliders. He even once survived a crash in his glider and walked home from the crash site despite several broken ribs and bones. Bud also deeply loved his avocado farm in Southern California and would often bring a fresh batch of avocados to meetings. The CNES people were somewhat suspicious of Bud as he was seen as a 'watchdog' but eventually understood that his role was to improve outcomes. Bud would attend many of the joint meetings before I left the US and later would spend a lot of time with me in France on the occasional quarterly visit.

My arrival in France with a young family came in late March. We arrived in Toulouse, and one of the French team, Christian Sirmain, greeted us at the airport with flowers. As it turned out, we didn't have the proper visas and working papers but came as tourists on advice from CNES. This was back when immigration rules were relatively lax, and we could get away with such things. Once Christian dropped us off at our new apartment complex known as FIAS, we unpacked and headed out to buy food at the local store. We had no car and had to walk to the store and carry home bags of food. It was good exercise, but I felt so utterly alone in the world at this point that it's still hard to describe. I spoke no French and knew little about the country when I arrived. I had been too busy finishing my thesis and graduate studies to learn much about my new home. It was a pity, really, but I had audited a short semester of French in graduate school and at least recognized some words. I had always wanted to learn foreign languages, but the cruelty of life was that once I had a real chance to use it, I had no time to learn it.

Once I arrived in France that cold spring afternoon, I knew it was completely up to me to provide for my family and make my way in this completely foreign land. At this point in my life, I suddenly realized how crucial cultural familiarity was and how important understanding the language was to navigate even the most basic transactions in life. Even the simplest things, like buying milk, were a puzzle. On that first day, I began solving a lifetime full of mysteries. This ability to push aside fears and self-doubt, observe and analyze, and finally decide would become one of the most valuable skills I would develop. It all started here, or so it seemed at the time. Within a few days of my arrival, I began to understand that there was still so much to learn in this world, and I was starting that adventure that would never really end.

The apartment complex at FIAS was an interesting place. It was built next to the CNES facility so I could easily walk to and from work. It also housed adult families and people from all over the world who were in France to work at scientific institutions, attend French technical universities, or receive training as fighter pilots in French Mirage planes. I was amazed that our neighbors included Brazilians, Iraqis, Egyptians, Poles, Syrians, and Australians. This international melting pot made my experience at Caltech's Braun Hall pale in comparison. The smells in the hallways went well beyond the curry that I so strongly remembered from Caltech, and the languages varied from English to Arabic to Italian. The apartments were equally foreign, with one bedroom for two adults and one child and a tiny kitchen with an even smaller refrigerator.

My first day on the job came on a Monday. Nobody told me when work started, so I showed up at 0700 at the main gate to CNES. While this was several hours before anyone arrived, I was at least on the access list and was greeted with less suspicion once I showed my passport. One guard spoke English and explained to me that nobody showed up much before 0900 and that I would be better served to take some coffee and coming back later. When I returned later at 0900, I found my way over to the building where I was working and found my boss's office. My direct boss would not be Tarrieu but instead would be Henri LaPlace. Henri was a physically robust older gentleman with a full facial beard and a gentle manner. He smiled widely and firmly shook my hand as I entered the office. Henri was an old hand at CNES and had spent most of his career there. His history with the Soviets and Soviet cooperation went back at least 20 years to when de Gaulle made his 1966 trip to Baikonur, deep in the Soviet steppe.

Henri was with him on that trip, a historic occasion during the Cold War. Perhaps one of the most notable items on de Gaulle's agenda focused on Soviet-French cooperation in space. When de

Gaulle visited the Baikonur Cosmodrome, he was the first Western leader to do so, and with Soviet leader Leonid Brezhnev, they watched the launch of the Kosmos 122 satellite. This made de Gaulle the first Westerner to see a Soviet rocket launch. This gesture preceded an agreement signed by the two countries for peaceful space exploration. Decades of cooperation in space research followed, including the launch in 1982 of the first French astronaut Jean-Loup Chretien on a Soyuz T-6, during which a series of Soviet-French experiments were performed. Like my earlier experience at JPL with Jim Burke and Bud Schurmeier, I was stepping directly into the history of the Space Race with the Soviets but now on the other side of the Atlantic.

Author's office in France at the Centre Nationale D'Etudes Spatiales. Circa 1990. Image Credit Jim Cantrell

Henri and I spoke for about 30 minutes in English about the job and his expectations of what I would do. We then began discussing the details of the Mars Balloon design. After about three or four sentences in English, he looked up at me and said in his passable but highly accented English, "This is the last we speak English. From now on, we speak French as this is bullshit". I nodded in agreement and began listening to him speak a foreign language very slowly and show me graphs as he went along. I was learning about their Mars

Balloon designs and a new language simultaneously. I would look back in my mind's eye at this moment months and years later. This was a turning point where I truly became part of another culture and another world. Henri's insistence that I learn French, as imperfect as it was, and his constant corrections and encouragements made a massive difference in my ability to assimilate with the team. Most Mars Balloon team members spoke English, but the quality varied. I eventually found the challenge of a new language very rewarding and made efforts to listen and learn. As six months passed, I was functional in French, and after twelve months, I was becoming fluent. It was only after two years that I could have conversations on the phone which lacked the visual context of in-person conversations. With this, I honestly considered myself to have mastered the language.

The first months on the job went very quickly, and I spent a lot of time trying to understand exactly what the Mars SNAKE design would look like. CNES had decided to make the long structure into a sub-surface radar antenna. The SNAKE was now looking to be 30 feet in length with another 30-foot long 'tail' that dampened tail dynamics as it dragged across the surface, and it made a perfect long-wavelength radar to search for water underneath the Martian surface. The electronics would be housed inside the SNAKE and drive the outer shell as an antenna.

CNES hired a young engineer named Michel Moreliere to build the electronics for the ground-penetrating radar. Michel had just returned from a stint at ESA and had a young company in Toulouse where he would design and manufacture radar hardware. Michel was a very charismatic man about my age and had an excellent command of English. He was highly driven, liked to build things and soon became a very good friend. We spent a lot of time professionally and personally during this time and have maintained a friendship ever since.

As I was designing the SNAKE, one of the main challenges was the weight of the entire structure (less than 10 lbs. but 30 feet in length) combined with the frigid temperatures. Each tapered segment of the SNAKE was made of very thin titanium and had to be articulated together to form a flexible yet durable structure that could survive dragging along the surface for 1000 km on Mars. The joints between each segment had to operate in the presence of Martian dust and at -200F temperatures. To state that it was a challenge to design is an understatement. Christian Sirmain was our systems engineer on the project, and he took an unusual and irritating interest in the SNAKE's mechanical design, especially the articulating joints. True to my roots growing up on a farm and building go-carts from scratch, my design used a simple cable flex joint mounted to two end caps with the correct geometry. This made the part reliable, inexpensive, and rugged. Christian thought this design was too pedestrian for reasons he could not articulate and instead wanted to embark on a much more complex system. Christian and I fought these battles every Monday morning during the project meetings, and I think the team liked watching us spar. One thing is for sure, my time in France taught me to be aggressive and fight for my ideas, as the European engineering culture was based on ideas with the most merit being fought over the hardest. This could not have been more different than the U.S. engineering culture, which was far more collegial, and discussions were usually polite and deliberative.

My friend Michel took pity on me and offered to take me for a ride in his airplane and later tour some model aircraft he and his father built. Saturday began with an early morning aircraft ride and a visit to his father's house. Michel's father was a practical man; you could see it in his eyes and feel it when he shook your hand. He smiled when we asked if we could see his place where he was building the model aircraft and took us to his basement, which incidentally was a rarity in France. The door opened into a vast cavern filled with machine tools, model airplanes, well-organized tool cribs, and benches where several aircraft were assembled. While he toured us around the shop and as he was explaining the exquisite machining that he had done, this gave me the idea to build a prototype of the SNAKE joint and segments. I brought the idea up, discussed it for about 30 minutes, and had an agreement to start on it in the morning. We worked over the week, and by the time the following Sunday rolled around, we had a working prototype of the Mars SNAKE. I brought this into the Monday morning project meeting and placed it in front of Tarrieu. He smiled, moved the segments around, and then turned to Sirmain and said, 'It looks like the American was right.' I was vindicated, but the lesson was not lost on me that I would have to crawl out of my self-imposed shell and fight for my ideas as I could rely on a few others to do that for me.

The summer of 1990 was to be a busy one. We had numerous balloon flight tests planned to validate the basic super pressure balloon, a free flight in the desert to test the SNAKE and balloon system in ground-contacting mode, and a trip to Russia planned to evaluate a SNAKE radar test site in Latvia. As is the tradition in France, nearly the entire country takes the whole month of August off. This idea of taking a month off work is both very foreign and appealing simultaneously. My time frames would not allow it, but I still needed to work around everyone else's schedules. We had planned a desert test in September in the Mojave Desert, so I planned to spend my time back in the U.S. after the test catching up with family. Still, we had a lot of preparation for the tests, and I had the entire country of France to myself during August and remained at work.

Our first balloon flight tests occurred out of Gap, high in the French Alps near the Italian border. This small town permitted us to fly the balloon at 120,000 feet in the atmosphere, where the conditions resemble the Mars atmosphere, and cut the balloon down west of the site near our other test area at Aire Sur L'Adour in western France. We brought several balloon test vehicles and gondolas that would transmit us position and critical balloon performance data. They also accepted commands to deflate the balloon and cause the gondola payload to parachute down to the ground. It was customary to place identifying plaques on the payloads in the rare case that we lost the payloads, and, in the hope, this would encourage anyone who found it to contact us. In our rush to leave the facilities in Toulouse and make it to Gap, we forgot to mount the identification plaques. As it turns out, I was the only one who brought business cards to the test, so we decided to put my business cards on the outside of the payloads.

Our first balloon, consisting of some 5000 cubic meters filled with helium, lifted off from Gap early in July. The ascent was uneventful, and everything appeared to be performing nominally. During this time of the year, known as the stratospheric turnaround, the balloons would drift overhead for some time and then head slowly west due to the easterly zonal winds at 120,000 feet. This went as planned on the first flight, but when we went to cut the balloon down, we lost communication with the gondola. This would typically be a good reason for panic, but in this case, the balloon was heading for the open Atlantic and would be expected to be well above any airline traffic. We could not track it in any case, so we could only report this as a rogue balloon to French air traffic controllers. As I discovered later in my career, a 'rogue balloon' business was a huge deal to the U.S. Federal Aviation Administration and would revoke your license to fly balloons. In France, the attitude was much more casual and without any noticeable consequences.

Our next balloon flight used a backup of the same equipment and was launched a few days later from Gap. This one performed as expected and remained close to our target drop location. Unlike the last balloon, when we commanded the gondola to terminate flight, it separated from the balloon and headed down toward the recovery point. The balloon, however, did not self-destruct in the way it was designed and managed to stay aloft. The balloon had a hot wire in the bottom of the balloon fabric that melted the thin Mylar, and the deployment of the gondola would tip the balloon upside-down encouraging the helium to empty from the balloon. This did not work as planned. We monitored it for a few hours as it dropped to about 60,000 feet and wandered north as it encountered the lower atmosphere circulation. Again, we notified the French aviation authorities, and nobody seemed overly excited about this latest failure.

The next day, I took the opportunity to head back to CNES with a colleague who needed to return early and return to work. I was alone in the project office when the phone rang, and our secretary answered. It was from a Colonel in the French Air Force looking to talk to someone about the rogue balloon we had just launched. It was menacing the Paris air space floating at 40,000-50,000 feet. It was still too high to interfere with much air traffic, but the angst among aviation officials was high enough that they notified the French Air Force (otherwise known as L'armee de L'air). Annie, our secretary, convinced me to talk to the Colonel, and I brought my best French. By this time, my French skills were beginning to be passable, but most of what I had learned was from hanging around in the CNES machine shop and talking with the machinists. To say that most of my French vocabulary was not meant for polite society would be an understatement. As I spoke on the phone, the limits of my French became immediately apparent, and we switched to English. I was surprised to find out that this gentleman wanted advice on how to shoot down the balloon with military aircraft. He informed me that they had sent Mirage fighters to do a fast pass of the balloon on a vertical trajectory, which looked primarily intact. He asked me about using large-caliber bullets and air-to-air missiles to cause the balloon to burst. After a 30-minute-long surreal discussion about shooting down our balloon over one of the most beautiful cities in the world, we parted company. I never knew if they tried to shoot the balloon or let it drift away.

Our second balloon came back to revisit me later in August. I worked alone in the office again during the month when everyone else would be on vacation. I enjoyed the solitude and the ability to do much without interruptions. Many preparations were going on in the U.S. for our balloon tests in the desert later in September, and this also gave me more flexible hours to work with my colleagues there in the U.S., eight hours behind us. I was beginning to get used to the rhythm of working quietly at CNES during the day and spending the evenings on the phone at home discussing progress with my U.S. colleagues. When the phone rang in my office late in the afternoon that day, I assumed that it was one of my colleagues from USU working on the SNAKE prototype that we would be testing in a few weeks. When I picked up the phone, the voice was American, but this was a Sheriff from Cook County, Illinois, near Chicago. He was looking to speak with me and explained that he had come across a rather mysterious object with my business card and was trying to figure out how to return it to its rightful owner. As the Sheriff described it, he was called to investigate a large clear plastic bag draped over utility lines on a county road. Once they successfully removed it from the lines, they found the gondola still attached to the balloon and my business card. I was somewhat horrified but simultaneously amused that the balloon had made it this far yet was found. This was our rogue balloon that disappeared from GAP. The Sheriff mentioned that since this was a French government device, he would box it up and hand it off to the French consulate in Chicago. Some months later, we would receive this box, stern words, and warnings from the embassy officials about not repeating this.

It was a hot Saturday morning, August third, and I headed to the local Press stand in the town of Castanet where I moved to in July. I constantly searched for the International Herald Tribune (IHT) in English for the day's news. The IHT was published by the New York Times and included a lot of other English-based reporting, and I depended on the Saturday/Sunday edition to find out what was happening in the world. I was shocked to see that on this beautiful Saturday, Iraq had invaded Kuwait. I did not appreciate the enormity of this single act and the forcefulness that the U.S. would respond with. Oddly enough, when I lived in the FIAS apartments, I met several Iraqi families and military men in France to train in Mirage jet fighters. I had to imagine they were all headed back to Iraq to join the war.

Meanwhile, my colleagues at Utah State University were busy working with The Planetary Society to build a new generation SNAKE and instrument it for testing in the Mojave Desert that September. They constructed an engineering model of the device out of stainless steel instead of titanium and had it ready for testing by late August. Following the annual vacations, we packed our balloons and launching gear and headed to the U.S. This was a rather sizeable logistical chain finally arriving at Los Angeles International Airport. Lou's team helped get it through customs and arranged for all the on-the-ground logistics for the tests. Lou had gathered experienced volunteers (many were retired JPL engineers) and stretched members' twenty-five dollar checks to support a real mission to Mars. This was a remarkable achievement and evidence that Mars exploration was no longer the sole domain of nation-states. It came down to a few determined souls and a modest amount of money to make the critical contributions.

Just like a few years before, our team gathered in the same Caltech conference room to review the test plans. We had personnel from JPL, CNES, IKI, Lavotchkin, and Caltech. The transportation and chase vehicles were from Caltech Geology, courtesy again of

Bruce Murray running the Geology department, and some private cars were also brought along to complete the testing logistics train. Our goal was to test the combined Balloon and SNAKE in a free-flying configuration and to gather environmental data inside the SNAKE on Mars-like terrains. Jim Burke had spent the summer scouting spots in the desert for the testing by flying his Cessna over candidate areas. Jim's father was a WWI ace fighter pilot but had fallen on hard times during the 1930s and the Depression of the era. The family took to living in the Mojave Desert, and Jim knew it very well from sight. In his aerial wanderings, Jim had found two perfect sites: one made of large dunes and one of combined rugged lava and sand with little vegetation. Both sites were near our original testing sites at Lavic Lake.

As we departed Pasadena in early September, I was assigned two Russian VIPs from NPO Lavotchkin: Konstantin Pitchkadze and Gary Rogovsky. Kosta, as we called Pitchkadze, would go on to replace Kremnev after the failed Soviet hardliner coup just a year away and would retire as the General Director of NPO Lavotchkin. Lavotchkin, as we called it, was a Soviet-era manufacturer of aerospace products, much like the American Boeing corporation or Lockheed Martin. The company was founded in 1937 as OKB-301, a Soviet aircraft design bureau (OKB), and manufactured piston-powered fighter aircraft that led the defense against the German onslaught of 1940 and 1941. In 1945, Semyon Lavochkin was promoted to the head designer of the design bureau, and its name changed to NPO Lavotchkin. Later, it was responsible for all Soviet interplanetary probes, including the Luna sample return program, the Soviet Mars probes, the Lunokhod program, the Venus Vega program, and the Phobos program. Ironically the Nazi army almost captured the aircraft plant in Khimki, Russia, where I would later spend extended periods working on various space missions.

Kosta was a pleasant man of Georgian origin, which meant he had a distinctly different take on the Soviet way of doing things and more openness to foreigners. He had a large frame and a broad and friendly smile. I immediately liked him. Gary was his right-hand man in most programs they undertook and had a distinctly military air about him. He strangely reminded me of my grandfather, which colored my perceptions of him. Gary and I would later become very good friends and share several adventures in common before his untimely passing in 2009. Neither Gary nor Kosta spoke English well, and my Russian skills were nonexistent at the time beyond keywords like 'beer' and 'toilet.'

As Bruce clearly and sternly pointed out, my job was to make sure that our Soviet guests were fed, made it to the hotel, and made it to and from the tests over the next week. My U.S. partner for this was Craig Christensen, who became involved in the Mars Balloon program as a graduate student at USU. It was about noon when we left Pasadena, and we decided to eat lunch. I tried to ask my Soviet guests if they were hungry and quickly realized the futility of my efforts. I took to sign language and signed to them the act of eating. Their response was an enthusiastic 'да!'. I didn't bother to ask them what kind of food they preferred; instead, I decided to treat them the way they treated us when we visited the Soviet Union. Just take them to eat.

We arrived at my favorite Chinese restaurant on Colorado Blvd in Pasadena. As we sat down, Craig sighed and looked at me inquisitively. 'How are we going to get their orders and translate from Mandarin to Russian?' he asked. I started to ask questions of our Soviet guests only to be rewarded with long phrases in Russian that I could not understand. Failing at that, I tried my best Russian and asked if they wanted some beer. We got an enthusiastic 'да' and Craig ordered a bevy of dishes that we all shared. Suddenly we were feeling more like diplomats than pioneering engineers. When we finished lunch, our Soviet guests thanked us, and Kosta uttered out a few words in English 'You are now our mother'. He repeated it in Russian just in case we understood 'ты наша мама'. He then smiled and gave me a big hug. I was now a parent of two Soviet space officials and had a week to ensure no harm came to them..

We arrived in the early evening in Barstow and met up for dinner with the rest of our colleagues. Plans for testing the next day were discussed, and times were set for heading out. As we ate, storm clouds were forming to the east over the desert, and we faced the possibility of a stormy and wet test area in the morning. The decision was still made to proceed at 0300 in the morning and head to the test site near Lavic Lake off Interstate 40. This first test would be straightforward, involving inflating the balloon, attaching a gondola and SNAKE, and letting it drift across a dry lakebed. The goal was to test the basic functionality of this strange flying contraption that we had built in anticipation of more rigorous testing over the next few days.

It's always amazing how short the night is when you depart the hotel at 0300 in the morning. I have always liked the early mornings, so this was not such a burden. As I collected my Soviet colleagues from their hotel rooms, I could tell they did not share my enthusiasm for the early part of the day. Nonetheless, they obediently accompanied us to the test site. They sat in the very back of the Suburban and didn't talk much, if at all. We arrived at the appointed site after an hour's drive. After a few cups of coffee, I was ready for the day. We arrived on the flat, dry lakebed surface as the crew set up for the balloon inflation and flight. We planned to inflate and fly as the sun rose on the playa.

One of the problems with such testing, as we discovered that morning, was that nature typically calls following the consumption of coffee. Fortunately, if you are a man, all the world is your restroom. This one, however, called for something of a more primitive nature, and I went searching for some toilet paper and a shovel. My grandfather had taught me well by camping when I was a very young lad and now that skill was paying dividends. I headed as far away from the camp as I could go and behind some bushes. Once I constructed an appropriate hole, I took care of business.

What I had yet to anticipate, however, was the presence of desert creatures, especially the diamondback rattlesnakes, that inhabited this area. One advantage of these species over humans is their ability to sense the vibration of our walk and see body heat like an infrared camera does. I had no clue that the diamondback was curled up in the bush six feet from me. I suppose he hoped I would continue and not stop in his vicinity. Despite their reputation, rattlesnakes do not like humans and generally are frightened by them. All this logic was worth nothing as he began rattling rapidly as I was in the most vulnerable position I could be. Fortunately, his fear of me over-rode my fear of him, and he ran off into the desert with rattles blazing before I did the same with likely unpleasant results.

Once I finished, Kosta, in his broken English, was asking me about bathrooms. I handed him the shovel and the toilet paper and pointed him in a direction where I didn't think rattlesnakes would be lying in wait. I didn't know how to explain the snakes nor have the language skills to explain it, so I sent Kosta, my responsibility, into the desert to accomplish his business. I was glad it was all dark, so we were all afforded some privacy!

As the sun rose over the playa, we began inflating the balloon as the flight train operated as expected. We had a mobile van assembled with the telemetry receiving equipment and logging data from the gondola and the SNAKE. What started as a roar coming from the sun's direction suddenly became a loud shriek as an F-18 moving at an estimated 500 MPH at about 1000 feet off the deck passed just above us. I hit the deck out of instinctive reaction as the jet passed too late for practical protection, but the balloon rocked back and forth in the jet's wake. The Marine pilot from Twentynine Palms Marine Base took the F-18 vertical and hit the afterburners as he went nearly vertical for about a minute. We all stood at attention watching this spectacle, and few words were spoken. This was the sound of freedom, and I wondered secretly if this display was meant for our Soviet colleagues as Russian could be heard spoken on the open radio frequencies. It was an impressive sight, and we were happy that they didn't return for a second look.

The balloon test went off without a blemish. The dawn winds were very light enough to gently push this odd-shaped alien contraption along the flat sandy surface of the dry lakebed. The balloon system moved along with an animal-like gait where the SNAKE would advance, stop, and then move a few feet forward. This causes the balloon and gondola to act like a pendulum of a grandfather clock and lurch back and forth in a slow rhythmic motion that somehow seems alive and well-coordinated. In the back of my mind, I wondered what such a thing might look like on Mars and if we did succeed in getting it there, if some alien civilization might somehow find it, and wonder themselves if this was a mechanical creature of some sort sent to an alien world to explore it. We followed the balloon system in the Caltech Suburbans as they slowly traversed the playa. Several cars were filming it, and it was like some pre-Burning Man parade of bizarre machines in the desert. My Soviet colleagues jumped into the Suburban back seats again and watched through open windows as we witnessed the very first terrestrial traverse of our Mars Balloon. Significantly few people were without a smile that morning.

The Mars SNAKE in action on an ancient lakebed. Image Credit Jim Cantrell

The transparent cylindrical balloon was about thirty feet long and five feet in diameter and was about 1/100th the size of the balloon that would fly on Mars. It was much smaller here on Earth since our atmosphere is about 100 times denser than Mars. That worked in our favor for this testing as even the slightest winds will result in a significant force on the balloon film and anything attached to it. One would assume that a Mylar bag filled with helium would be a simple proposition. Still, these tests showed us all that there was much to learn about deploying, inflating, and flying such thin helium-filled membranes and the flight train that would interact with the irregular planet's surface. We didn't speak about how we would package such a system up into a probe the size of a washing machine, fly it for two years to Mars, slow it down from 25,000 MPH to a speed where we could deploy the 3.5-micron thick Mylar film at -250F on Mars and then inflate it in less than 2 minutes. That was what we were still working on back in Toulouse, and for now, getting terrestrial experience with these devices informed our designs that would later make it to Mars.

After the tests, we packed up and headed off to the hot playa, where temperatures exceeded 100F by ten in the morning. We headed to another hotel deeper in the Mojave, closer to the following test site, and retired to lunch and a pool. Fortunately, I discovered Denny's restaurant nearby, where we stopped for lunch. The key feature I was looking for at Denny's was their menus with pictures of the meal. This significantly reduced the language barriers for my Soviet colleagues. After a "point and eat" lunch, we bought beer and headed for the hotel pool. By the time dinner came around and I went to gather my Soviet friends, they were still in the swimming pool and had absorbed enough sun on their Moscow-whitened skin to give me a serious worry about sunburns. I motioned to them to come and eat by mimicking eating with my hand to mouth. Gary Rogowski then blurted out his hitherto unknown English vocabulary and said, 'No more eat. We sleep'. I mimicked an OK symbol and then held up three finger symbols and exercised my budding Russian skills 'zaftra ootrum' - meaning tomorrow morning. They nodded, and I smiled and headed for dinner.

Our next test would be more ambitious, and the test site would have lava and more severe obstructions. This would test the dynamics of the flight terrain on rough terrain and the ability of the SNAKE to avoid snagging in the terrain with realistic flight dynamics associated with the balloon. We planned to inflate it like the first test at dawn and let it go by itself for a few miles downrange. I was to be in one of two Suburbans downrange from the inflation site, waiting to capture the flight train. While we didn't discuss how we would capture it, I had enough confidence that it would not get past me. Kosta and Gary stayed behind at the launch site, and Craig and I manned the

Suburban downrange in wait for the balloon. We kept in touch with two-way radios and knew the moment that the balloon was inflated and then released.

We watched as the stiff winds blew the Mars Balloon over the rugged lava terrain. The balloon made the same motions as the day before, moving back and forth, but this time, the vertical motions of the balloon were more pronounced. We watched through binoculars from about 2 miles away and saw a crowd of people following the balloon alongside cameras and others just as spectators. I noted that they were running and estimated a speed of 5-10 miles per hour on the lava. It would be only 15 minutes or so before the balloon was upon us. We estimated the trajectory of the free-flying Mars Balloon and attempted to line up with it. Fortunately, the winds were cooperating and drifted mainly straight toward us. There was also a chase plane overhead - a Kit Fox plane piloted by one of the TPS volunteers. Someone in the plane shot a video of the balloon moving across the lava. Once it got closer, we could see that the SNAKE was traveling effortlessly across the lava surface and quickly crossing three- to four-foot-wide chasms in the surface. The spectators, who included Kosta and Gary, ran alongside but needed help navigating the lava than the SNAKE was. We had the same trouble in the Suburbans, and I tried to consider the possibility of a harrowing chase in the lava. Fortunately, we found a crossroad through the terrain almost perpendicular to the balloon's path and could easily intercept it if we were good at anticipating the crossing point.

As the balloon got within 200 feet of us, we could see the shimmer of sunlight off the SNAKE's outer surface and its fluid motion over the rough terrain. Again, the Mars Balloon flight train looked like an alien life form running away from its human captors. Craig and I prepared to capture it and aligned with it on foot as it approached us. Bud Schurmeier showed up in his off-road Volkswagen beetle converted into a sand machine at the last minute. He got out to help with the capture. As the balloon approached us, we could see Lou Friedman and Gary jogging behind it, red from the heat and jogging and nearly out of breath. We snagged the balloon by grabbing its tether to the gondola and recovered it without incident.

None of us had to say that morning that the test was enormously successful and that we had achieved something significant.

The last test was to take place on a dune. We had decided that this was the third class of terrain we might encounter on Mars, which posed some unique dynamic challenges to the flight train. Jim Burke had located what he thought would be the perfect place for the test just outside of Desert Center, California. This lonely town is along Interstate 10 between Blythe, California, and Palm Springs. It's also the modern-day site of one of my favorite racetracks, Chuckwalla Raceway. It was a good two-hour drive from the Barstow area, and we decided we needed to head there early in the morning. We woke up at 0230 and headed out again at 0300, hoping to arrive at 0500 at the test site. I gathered my Russian colleagues from their hotel room filled with beer bottles from the night before. They spoke little as we drove silently in the dark down a very narrow two-lane desert road. Craig took command of the other Suburban, and we caravanned across the desert at this ungodly hour.

Given the hour, the lack of anyone on the road, and the extra time we might need to catch up with the others who had left before us, we would be well served to enhance our traveling speed. I discovered that the Chevrolet Suburbans did not have a speed governor but a practical top speed of 110 MPH. Craig kept up with me as I accelerated down the deserted roads. My Russian colleagues seemed to be enjoying the fast pace. As we passed over dips in the roads and experienced some slightly negative g forces, they surprised me with a yell out in unison of 'oooplahhh!' followed by childlike laughter. Once the road straightened out, I turned on the radio in time to find a station playing an apropos song: Bill Warnes and Jennifer Warnes singing 'The Time Of My Life". Yes, I could not agree more.

Despite our late start, we arrived at the rendezvous point to the dunes at the same time as the others. Nobody spoke of my probable high-speed transit, and I was happy not to be questioned. We assessed that the dunes were easy to reach but would need to be careful with the two-wheel-drive vehicles as we crossed a long sand field. We headed out to the dunes seen in the early morning light about 4 miles away. Once we arrived, we assessed the likely wind patterns and located a place to assemble the flight train. Once again, the usual buzz of activity began as the French and American teams began assembling all system elements seamlessly.

About 30 minutes into this activity, I spotted a truck approaching us across the same sandy plain we had just crossed. Looking through my binoculars, I could see that this was a federal vehicle of some sort, a National Park Ranger it appeared to be. As he got closer, Lou Friedman and I walked out to greet him as to avoid questions associated with the presence of the Soviets and French. The Ranger got out of the vehicle and explained that he was a Park Ranger and that this was federal land. He asked us more than a few questions about what we were doing and if we had a permit. Hearing three languages and seeing a strange object assembled in the desert might get us arrested in 2019 on terrorism charges. In 1990, however, it was taken more in stride. The Ranger explained that this was a historically significant and sensitive site, 'early man' as he described it and that we could not use it for testing. We understood and agreed to move. The Ranger helped suggest a similar site 'just a few miles away' in absolute distance but, unfortunately, about an hour's drive to reach by highway.

Time pressures forced us to consider driving across the desert. Our intrepid Kit Fox pilot had landed his plane on the sand like he had done on a narrow strip of sand in the lava field the day before. He offered to take a radio and act as the pathfinder for the overland route. We took this course of action and headed north into the desert. I was thrilled that we had four-wheel-drive vehicles and had thought enough to bring tow chains, winches, and spare tires. We needed them all as it turned out. It took us about two hours to cross this short distance, and about halfway along, we decided there were better ideas than this.

When we arrived at the designated alternative site, vectored in by our bush pilot in the Kit Fox, it was beginning to pass 100 degrees in the shade. Despite that, we were determined to make this test happen on this day, and this was a perfect spot with several miles of dunes to traverse and a favorable wind direction for the traverse. We quickly unpacked our gear and began the process once again of assembling the flight train. We were ready for balloon inflation at about three in the afternoon. Our Soviet colleagues were already hitting the beer, and the French team was swigging on a cold bottle of white wine. I stuck to the water as I oversaw the recovery crew again. I headed out with Craig in the other Suburban and Bud in his VW bug to find a path to a logical recovery point around the dunes. Gary and Kosta decided to join me in my Suburban and again got into the back third seat laughing and shouting 'oopla.' I meant they enjoyed my wild drive early in the morning and sought more.

We were unsuccessful in finding a way around the dunes and decided among the three of us that the only way through was up and over the dunes. These dunes were not exceptionally soft sand, but they presented a challenge on height, around thirty feet tall, and the slopes, which were very steep on the lee side. I started slowly up the first dune and got stuck. I backed down in a four-wheel drive and regrouped. We all decided that the best way to approach this was with speed, as Bud and I had some experience with dune driving. I backed the much-abused Suburban up far enough to get a good run at the mound in four wheel drive and, to my pleasure, could drive straight up the dune and descend effortlessly down its backside. Thinking quickly, I knew I had to keep my speed up and keep my foot on the throttle as I approached the next dune with the same result. I could hear my Soviet friends bouncing around in the back as we attained zero gravity at the top of the dune and fell back into the seat as we descended off the dune. They kept crying 'Ooooplaaaah' and laughing hysterically. If I had not been so concentrated on making it over the next dune, I would have joined them in laughing.

I stopped for a breather once we arrived at the point where the dunes stopped. I watched as Craig and Bud followed my approach over the dunes and slowly and softly glided these multi-ton machines softly down the dune. It was like watching a ballet. When they got out, they laughed nearly as hard as Gary and Kosta. We drank some water, regrouped, and checked in with the balloon team. They were ready for inflation. We decided to split up and cover a more expansive line of interception parallel to the anticipated line of travel. Once in place, I gave the go-ahead to release the balloon. By this time the wind at our location had increased considerably since we were approaching later afternoon and the sun was starting to descend in the sky. We could not see the balloon, so it was a very long wait at the recovery point, not knowing where it was heading. We monitored the radios and knew the balloon was doing well and outpacing those on foot. For a glorious 15 minutes, the balloon was flying as it was meant to be and under no control. This was the best test possible.

We began to see the balloon popping up and down from behind the dunes. With each successive dune, its apparent altitude got higher and higher. The three vehicles at the rendezvous point agreed to collect near the apparent exit point from the dunes. Behind us was a large stand of desert Palo Verde trees covered with thorns. We also worried about snakes coming out this late afternoon looking for rodent prey as the sun went down. We could see numerous sidewinder tracks in the sand, which only added to our overall anxiety.

As the balloon exited the dunes, it looked like it was going at least 50 miles per hour. It was likely only moving half that speed or less, but it was fast enough to give us pause grabbing it by hand. Fortunately, we didn't have to. I watched with amazement as the balloon ran directly into one of the Palo Verde trees, which towered some 20 feet off the desert floor. To our surprise, it slowed down but was not enough to stop the forward movement, and the snake climbed over the tree. My face must have looked shocked as it glided right by me, and I didn't even think to try and stop it. Fortunately, Craig was behind me and could grab it as it tried to mount a second Palo Verde tree. We had safely recovered the system, but it went far from planned.

I had lost track of Kosta and Gary, but they had placed themselves safely out of harm's way. Bud showed up around when Craig and I decided we needed to pull down the balloon. One of the graduate students from UCLA that had volunteered with Lou offered to climb the tree and pull it loose. We let him, but in so doing, the balloon was cut free and headed off into the sky. Given my prior two experiences with rogue Mars Balloons flying through congested airspace, I was horrified. We could do nothing at this point but watch it ascend. I radioed back to the launch base, and the French attitude was typically French. No big deal. I imagined that the balloon would be another unusual site for pilots traversing the desert southwest and hoped they could successfully steer around it.

We returned to the launch site to finish up with the crew. This time, we found a paved road that could get us there. Our bush pilot was offering rides in the Kit Fox as we arrived. He would take off from the dunes and land between them. I took the opportunity to go up with him and fly around in this little hot rod of the sky. He offered me the stick for a short period as I had some flying experience already. It was exhilarating seeing the sun setting over the desert and knowing that we had conquered the terrains down there with our queerest of aerial vehicles - the Mars Balloon. After packing, the French team broke out the wine, and Lou offered the TPS-furnished beer. Gary and Kosta cut up their sausages that they always seem to keep close nearby, and we had a party on the dunes. Many toasts were made, and it seemed like we were on a path to Mars that could not be stopped. Mars was our oyster, and our balloon would find its pearls.

The Mars test team at the conclusion of the desert testing. Image Credit Jim Cantrell

Our return to France involved the hard work resulting from a successful test campaign, including data assimilation and redesigns based on lessons learned. Nobody doubted anymore the feasibility of this crazy flying machine we had proposed only a few years earlier. The intricate engineering problems remained, and we built flight-like prototypes and tested them as best we could back in Toulouse.

Michel and his crew were busy designing and building the radar, which used the SNAKE skin as an antenna. We made a trip to Riga, Latvia, that fall, where we met with the radar team, who provided test support and some transmitters for the gondola. We organized a test on their famous dunes where we could easily measure the waterline and compare that with our probing radar return signals.

As I returned late that September to CNES, our project was in full swing, and life turned to a very intense but normal pace. My personal life also became more normal as I had become comfortable living in France, became fluent in the language, and began to venture out on weekends into the countryside to do some sightseeing. My son was born in January 1991, meaning he was now a French citizen, and I had learned how to navigate the French medical system.

As we progressed, I contemplated the inevitability of the mission launching in 1994 and making it to Mars successfully. There was much testing remaining and many engineering challenges to overcome. Bud Schurmeier continued to visit me quarterly, and the TPS team back in the U.S. was making significant contributions to the SNAKE design and prototype testing. It was hard to imagine what might come in the way of our success.

The war in Iraq brought unusual tension to our Mars Balloon project team. I watched in the winter of 1991 as the U.S. built up forces in the Middle East, obviously in anticipation of a large war. My French colleagues were primarily against this entire enterprise while I was still determining the whole thing. I had several friends involved directly in the war, including family. While I had misgivings about the entire enterprise, I didn't want to judge something I had no real insight into. From where I stood, the U.S. had built an immense arsenal to defend Western Europe against the Soviets, and the Soviet Union would likely pose no more threat to that region for the foreseeable future. In an unfortunate move for them, Iraq had invaded a neighboring country. It aggravated the same Western governments who now pointed this arsenal meant for the Soviets at Iraq. To me, the outcome was not in doubt. To my French colleagues, this seemed to be more of a matter of life and death, and they were confident that the American defeat would dwarf that in Vietnam.

I generally stayed out of politics in social or business discussions, but my French colleagues saw me as the sole representative of George Bush and the American government. As a result, I received a large part of their protestations on behalf of the French nation. As 1991 dawned, I watched the war unfold on television, and this was the first time I had ever experienced that. It was shocking to me to see the rapid violence of modern warfare, and the ability to watch it live made it very personal in a strange way. Watching it on French TV, I also learned a new vocabulary related to warfare and diplomacy. During those early months of 1991, we watched as the entire arsenal meant for the Soviet army coming through the Fulda Gap in Germany got dumped on the country of Iraq. The results were entirely predictable.

The spring of 1991 brought more meetings with the Soviet delegations as we continued to solve the challenging problems of packaging the balloon up to get to Mars and then unfolding it safely in the Martian atmosphere. For our bilateral meetings, we would almost always meet in Toulouse. Through these meetings, I discovered that the secret to the cooperation between France and the Soviet Union was these meetings in France where CNES would pay the Soviets a 'Per Diem' amount for each day that they stayed in France. These payments were intended to cover all lodging and means and were considered 'generous' by French standards. The Soviets would stay in the cheapest hotels and eat foods from the Soviet Union instead of the expensive fare in French restaurants. After initially considering this whole arrangement to be corrupt, I began to understand that it was a balance of forces like much else the French were doing.

Throughout my early days working with the Soviets, I began to appreciate their terrible situation and how it had been that way for centuries. In the late 1980s and early 1990s, we saw a clearer picture of the authentic way of life in the Soviet Union under despotic rule. When I was young, likely due to the Hollywood influence, I always considered the Soviets the equivalent of the superhuman 12-foot-tall race. We were taught to fear them and be prepared to defeat them should they come calling to over-run the United States one day. What I was finding in my observations could not be farther from the truth. The real dimension was sad and had a terrible human price. By our standards, the brilliant Soviet engineers and scientists we worked with were impoverished and destitute. Their country's infrastructure, what of it existed, was crumbling and was in disrepair. The real motivation behind these cooperative space programs with the French was these per diem payments and the immense personal prosperity they brought to their families. Seeing this was the first time I truly understood how genuinely fortunate we in the West were and how dark human existence can be under a totalitarian regime.

I took my first trip to the Soviet Union in early 1991 to Riga, where our colleagues at the Institute worked with us on the radar payload. They also provided communication systems between the SNAKE and the airborne gondola. The gondola would then send the data to the Mars Balloon Relay to be in orbit aboard the Mars Observer spacecraft. This relay, provided by the French Space Agency to JPL, would become the local global communication system around Mars for decades. This idea was a small side project developed by Jacques Blamont that summer in 1988 when I first worked at JPL. We thought that small, low-cost Mars missions operating on the surface needed something in orbit to relay data back to the Deep Space Network on Earth without requiring the sizeable table-sized dish antennas pointed very accurately at Earth. This 'AT&T' service around Mars was sold by Bruce Murray, Jim Burke, Lou Friedman, and me to JPL's Lab Director Lew Allen that summer. Lew, a storied Admiral and steady leader of JPL, approved this addition to the American satellite, and CNES provided the radio. Despite Mars Observer exploding on its way to Mars due to propulsion problems, the Mars Balloon Relay (MBR) led to many follow-on devices on Mars and is still in use today.

Landing in Riga in and of itself was an adventure as the regional airport also served as a Soviet Air Force base. We were warned not to take pictures as we landed or transited the airport. I looked through the windows and saw more partially dismantled and sidelined planes than operational fighters. Perhaps this was what they wanted to hide. We were met at the airport by our Soviet colleagues and taken to a nice hotel in the region of Jūrmala. This is nominally a beach resort, but this time of year, it was closed, and the weather was frigid. Nonetheless, we stayed there and could stroll on the frozen beaches. According to my French colleagues, this was high praise by our hosts for putting us here and was likely due to my presence as the first American to work with them. While the accommodations were sufficient, one could not help but notice how decrepit the infrastructure was and how meagerly the population ate. While none of us were in danger of dying of starvation, we did find the quantity and quality of food to be wanting. The facilities, in general, were old and in great need of maintenance. Some of the buildings we drove by on our way to the Institute in the morning were falling and would not have taken much of a push to tumble. Christian Tarrieu, in his most cynical humor, remarked on the way to the Institute that 'one of the greatest talents of the Russians was their ability to make a brand-new building look 100 years old the day after it is finished'.

I could deal with the crumbling exteriors, the sometimes-nonexistent roads, and the complete lack of pollution controls on the cars resulting in 1950s-like LA conditions of smog, but it was the bathrooms that I came to loathe the most. The bathrooms were by far below any standards that I had ever experienced. The Institute bathrooms featured open-style 'Turkish toilets,' which required the same skills we had developed in the early morning desert tests. Similarly, anyone who walked into the bathrooms would also freely see you 'doing your business,' which took a certain bravery. Some of my colleagues would refuse to use these facilities and wait until they reached the cleaner and more Western-built hotel bathrooms in the evening. One mental note I made was the lack of toilet paper, which was scarce and treated almost like gold in the Soviet Union. I, fortunately, had brought packets of Kleenex which I could carry in my pockets. Those small packages were my saving grace more than a few times. We started late in the morning most times, and I secretly imagined that this was to offset the timing for the use of bathrooms to favor the hotel bathrooms.

We worked late into the evenings after sharing a large lunch in classic Russian style with a table full of finger sandwiches, unusual mystery meat dishes, and vodka. Before entering the Soviet Union, I didn't care for the taste of vodka, but Russian-sourced vodka was quite excellent. Fortunately, God granted me a stomach made from cast iron, and I could handle almost anything that was fed to me. Fortunately, the Soviets set out to impress their first American guest, me, with the most unusual and, in their minds, highest-order delicacies as a sign of respect. I got introduced to 'kvass', a traditional Slavic beverage fermented from rye bread and for which I never developed a taste. I was also introduced to many cold fish dishes, many of which tasted delightful, but occasionally, a few smelled like the rotting Aral Sea. There were also many 'mystery meats' served on slices of bread like open-faced sandwiches; most tasted quite good. I found it to be good practice not to ask what the composition of the meat dishes was. All in all, there were many times when the vodka available to wash the cuisine down with was quite welcomed. I learned in these lunches to be gracious and eat what was in front of me. That sometimes resulted in less than spectacular evenings later in the hotel rooms as my body struggled to digest the food long after we finished the meal.

Our work with our colleagues went well and without anything notable or unusual. There was much discussion during social occasions of the political situation under Gorbachev and the hyperinflation of the currency there. That came into view one evening when we went into downtown Riga to go shopping. I had brought French Francs with me, and one of our guests brought along a gym bag full of Soviet Rubles to exchange as we needed to. The official exchange on the government market was two dollars per ruble, but the dollar traded on the street black market for 35 or 40 rubles per dollar. That evening, only a few of us were out on the town and planned to buy only a few gifts for our families back home. Yet, we had to bring a huge bag full of cash to exchange as inflation was so bad in the Soviet Union. We visited a few shops to look at typical Russian art and other tourist-type gifts. I immediately sensed how different shopping in the Soviet Union was. The first noticeable thing is that many shoppers were in the store, but few were buying.

The large stores were open in the middle with counters outside the room, much like a Western bakery. All items for sale were on the wall behind the counters or in glass cases that formed the counter. Most stores had no rhyme or reason for the items for sale or displayed. You could easily find a single comb for sale next to a set of three knives and on the wall behind the counter three or four Russian stacking dolls. Prospective buyers would ask to see the item and be allowed to examine them at the counter. Should they decide to buy, they then got a slip of paper from the shop assistant and were to go to the cashier. Once at the cashier, the total was added up on an abacus, an ancient machine used to count by moving stones vertically on strings representing digits and powers of ten. Once you paid the cashier, you could return to the counter once again with your receipt and receive your item. Typically, these items were wrapped in plain brown paper and taped together but rarely put in a bag. There must have been a national shortage of plastic shopping bags as everyone carried one around for these occasions. Often, they originated from Western Europe and included the store branding on them. Fortunately, our Russian shopping assistant had a few plastic shopping bags with him, and we could collect and hold our tourist items without incident.

After we had completed our shopping, our hosts recommended a dinner in Riga. Oddly, there were very few restaurants in the Soviet Union. These were usually state-owned and operated affairs, and one needed a reservation to ensure they would have food available for the guests. I wondered exactly where we would go and was surprised when one of our colleagues led us down a somewhat dark road in the center of Riga. As we passed the increasingly dark road, I questioned my wisdom in trusting our Soviet colleagues. Still, all my instincts remained optimistic that this was something I didn't understand rather than dangerous. I was thankfully proven right when our colleague located a large steel door on the side of what appeared to be an apartment building. He knocked on it, and about a minute later, the door opened only wide enough to allow a very young face to look through it. My Russian language skills were not very well developed at this point, but I could tell this was some form of negotiation. The door promptly shut, and we waited outside quietly for another five or ten minutes in the cold air, wondering what this all meant before the door opened again, and we could all head inside.

As we entered the door, we walked into a rather large room about the size of an auditorium full of people eating at long banquet tables and a band playing jazz on the stage. This must have been what a 1920's speakeasy was like as the mood here was joyous and the scene resembled a party more than a dinner. We sat down at a long table that had been prepared for us. It had numerous bottles of vodka placed every 3 feet on the table along with shot glasses and some small sandwiches to wash the vodka down with. As we sat down and installed ourselves at the table, our Russian guests began a long series of toasts which is a Russian national institution! The first toast was to our colleagues and how happy we were to work together. The second toast was to the country of France and its symbolism of "Egalite" and "Fraternite" (Equality and Brotherhood). There was a toast to me as the sole American in the group and to my lack of fear in visiting the Soviet Union. They also mentioned that the last American they toasted to was Francis Gary Powers. That didn't necessarily make me comfortable but it was good for a lot of laughs. One of the last toasts was to the 'Soviet Union and may it last long enough to launch our mission to Mars'. That one was funny but it had an edge of seriousness to it! After about fifteen such toasts, we finally got down to the business of eating and I could not have been happier.

The dinner and night out at Riga left me with an uneasy sense. I kept thinking of the famous song by Harry Chapin called "Dance Band On The Titanic." The words echoed in my head:

"I heard the dance band on the Titanic
Sing "Nearer, my God, to Thee"
The iceberg's on the starboard bow
Won't you dance with me
There's a wild-eyed boy in the Radio Shack
He's the last remaining guest
He was tappin' in a Morse code frenzy
Tappin' "Please God, S.O.S."
Jesus Christ can walk on the water
But a music man will drown
They say that Nero fiddled while Rome burned up
Well, I was strummin' as the ship go down"

As I listened to the band playing American jazz on the stage and saw people dancing, the metaphor was inescapable. This was the first time I imagined the unimaginable: the Soviet Union could collapse. This thought had never entered my mind, but somehow, I sensed during this first visit to the Soviet Union that this might be the last of it. Funny, these kinds of premonitions. This premonition would end up being about much more. Little did I realize I was on a collision course with the Soviet Union's demise, which would alter my career and life.

As I returned to France from this trip on Aeroflot, we flew from Riga to Stockholm and then to Toulouse. It was a long trip with plenty of time to reflect on everything. I could not shake the sense that something massive and historic was about to occur. In the meantime, we had much work to do on the balloon and the snake back in Toulouse. I settled back into my work routine rather quickly and attended to home life with my new son and his sister. I was set to go back to Riga at the end of August and would only think again about this trip much later.

CHAPTER 7

SOVIET CAPTIVITY

"Those who do not move, do not die.
But are they not already dead?"
- Jean Behra - Formula One driver.

The demise of the Soviet Union finally arrived on a cold December day in 1991. Boris Yeltsin, a primarily unknown politician until that year, became the first President of the Russian Republic after being handed the keys to the Russian nuclear forces by departing Soviet President Mikhail Gorbachev. The dissolution of the Soviet Union had been many years if not decades. The critical and pivotal moment when the reality of this impending change arrived for most of us came on the 20th of August 1991 when Soviet hardliners attempted a coup against Gorbachev. These 'putschists', as they were called, arrested Gorbachev and sent tanks into the streets of Moscow and the capitals of the Soviet Republics to let everyone know that the Soviet Union was here to stay, and new leaders were firmly in control. Radio stations around the country played the Soviet National Anthem on the state-run radio stations during these days of the Revolution. In Soviet times, communist block radios didn't have dials for tuning but three or four push buttons that only permitted listeners to tune into the stations the government masters deemed 'safe' to listen to. This kept out the BBC, Radio Free America, and other Western European stations that contributed to the 'cultural decay of proud Soviet citizens' by exposing them to rock and roll and other such forces of evil.

This hardliner coup did not happen in a vacuum. Instead, it came after nearly a decade of social unrest in the Soviet Republics alongside an increasingly strong call for republic autonomy and relief from decades of economic stagnation. Gorbachev tried to stem the tide of dissatisfaction by transforming the Soviet Union into a loose confederation of independent states but could not stem the massive social change enveloping him and the country. After a fierce week of tension and eventual capitulation, the hardliners stood down and were deposed when faced with the massive street protests of average Soviets against their rule. Most importantly, the stubborn reluctance of the Soviet military to open fire on its people made the putschists realize that this coup would not succeed without massive bloodshed, much like the original Revolution in 1918. History was indeed repeating itself. Boris Yeltsin, then a lesser-known politician in Russia, became a leader of the resistance by organizing political forces and making a symbolic speech standing atop a tank parked on the Duma steps. In the rest of the republics, massive protests against the Soviet reassertion of power challenged the coup, and much of that effort focused on the public government buildings and the radio stations.

Our trip to Riga that August was to discuss an upcoming test of the Mars Balloon in Latvia. We were arriving during a traditional August vacation time in both countries, but this had to do more with the test scheduling than anything else. None of us from CNES knew that we would end up in the middle of one of the most significant geopolitical crises of the 20th century. When I woke that Tuesday morning, I knew something was wrong. It was my habit to turn on the radio and listen to the morning discussions. Here it was in Russian and on Radio Riga. The Russians have a unique intellectual discussion in their media that somehow comes across as quite civil and thoughtful, unlike what I was used to seeing and hearing on CNN in the U.S. Instead of discussions like I heard during my first trip there, the Soviet National Anthem was playing in looped fashion over and over. I tried the other two stations available on the radio and got the same music. Something was wrong. I thought to myself that perhaps Gorbachev had died.

My French colleagues and I arrived for work late that morning at the Riga Technical Institute, with only one person showing up to greet us. We arrived to find our Soviet colleague alone in the meeting room. In a scene almost mocking the significance of the events we found ourselves caught up in and reminiscent of a Quentin Tarantino movie, our colleague played old Elvis Presley records on an old record machine with tears running down his cheeks. As we entered the room to this surreal scene, all of us were speechless. We stood there taking the scene in for what seemed like minutes before we asked him if he knew what was happening and where everyone at the Institute had gone.

We were then told that there had been a hardliner coup in Moscow and that many Soviet Republics had declared independence, including Latvia, where we were presently working. We were advised to return to France and that most of our colleagues had gone to downtown Riga to defend the radio station from Soviet tanks. We sat for a while longer in the room, discussing our options. We headed back to our hotel instead of following our friends into the heat of the battle that morning in downtown Riga wondering what we would do. While we contemplated returning to France, we could not contact anyone to change our flight reservations. Calling out of the country was not an option at that point. The hotel manager informed us that she was closing the hotel and that we would have to find accommodations elsewhere. We briefly considered taking a train to Helsinki and finding a way home, but every solution seemed more uncertain than simply waiting this out and heading back as planned in five days.

Someone in our group decided to see if the trains to Moscow were still operating, and much to our surprise, they were. Several of us packed up our things the following day and headed to Moscow on the trains to see what was going on there and hopefully find lodging. In the worst case, we had numerous colleagues who would undoubtedly put us up in their apartments. I suppose that we were either too young or too dumb at the time to be concerned for our mortality at that stage of life, but the adventure of heading into

Moscow to see if the reports of massive protests were right seemed irresistible.

In a near mirror of our situation unfolding in Riga, Latvia, my colleagues Lou Friedman, Tom Heinsheimer, and Bud Schurmeier had landed that morning in Anchorage, Alaska on their way to Kamchatka to perform some rover testing on remote terrain with our Soviet colleagues there. They arrived at the airport in Alaska before the last leg into the Soviet Union to the news of the Soviet Putsch. After much debate, they, too, decided that the ongoing strife was not much to worry about and to proceed as planned. Lou Friedman was no stranger to Russia and the strange politics that are encountered there, so it was no surprise to anyone when he declared the Soviet Putschists 'dilettantes' and paid them no mind. They headed into Kamchatka, performed the testing as planned, and soothed worried colleagues back in the U.S. who watched the drama unfolding in Moscow on CNN and other 24/7 news networks. On the other hand, we could not call back to our friends and family, who were worried but happy that we were seemingly not in the unfolding street violence and confrontations happening in Moscow. Little did they know that we had headed to Moscow to investigate things.

The significance of the Soviet dissolution was something few of us grasped at the time. The consequences of it have reverberated globally for decades with geopolitical repercussions that few of us imagined ahead of time or even during the uprising. Few of us, either professional Soviet watchers or casual observers such as myself, ever imagined that the Soviet Union would ever cease to exist, let alone with the speed that it dissolved. At the time of the putsch, none of it seemed to have any real significance, and the chances of the gigantic force of history known as the Soviet Union dissolving seemed very minimal, at the very least. I expected the forces of the state to, in short order, consolidate power and restore some semblance of order. That did not happen as it turned out.

Nothing seemed out of place when we arrived in Moscow at the Leningradsky railway terminal. There were, however, many people in the streets of Moscow, and this was abnormal for this time of year when most good Russians were out in their dachas, or country homes, for the month of August. This week they were all in the street, mostly walking around and looking at the tanks stationed around the city and the soldiers guarding them. Starry Arbaat Street, the tourist commercial center of the city, was operating with business as usual and lots of street vendors selling everything from matryoshka (stack dolls) to farm-produced vegetables. Walking closer to Red Square and the city center, we saw signs of the struggles from the night before. On Leningradsky Prospect, the 6-lane primary road in and out of the middle of the city, the public transportation buses were lined up crossing the street at one location and had been rammed and burned by Soviet tanks and troops. Armed soldiers patrolled the makeshift barricade but didn't seem intent on moving it. Citizens freely moved through and around the barricade as if nothing had happened. A few of us took pictures and took good notes of the situation. At this time, I began questioning the wisdom of coming to the city after all. There had been clashes between the military and street protestors and leftover wreckage.

We made it to Red Square, and apart from the massive numbers of people and armed soldiers protecting the Kremlin, there was nothing out of the ordinary there to see. We heard from a few people in the crowd that the tanks and the protests were mainly focused on the area around the Soviet Duma - or parliament building. This spot, later known as the Russian White House, was indeed where most of the focus was at by the time we arrived. This is where Boris Yeltsin famously mounted a tank parked on its steps and gave a defiant speech against the Soviet authorities who had ordered the coup. Approaching this area on foot took much work due to the density of people. As we walked along the Moscow River from the Kremlin, we could also see what appeared to be blockades of boats on the river tied together to prevent Soviet naval vessels from traversing the city. Whoever was organizing the resistance knew how the forces might try to control the city.

We made it to Red Square, and apart from the massive numbers of people and armed soldiers protecting the Kremlin, there was nothing out of the ordinary there to see. We heard from a few people in the crowd that the tanks and the protests were mainly focused on the area around the Soviet Duma - or parliament building. This spot, later known as the Russian White House, was where most of the focus was when we arrived. This is where Boris Yeltsin famously mounted a tank parked on its steps and gave a defiant speech against the Soviet authorities who had ordered the coup. Approaching this area on foot took much work due to the density of people. As we walked along the Moscow River from the Kremlin, we could also see what appeared to be blockades of boats on the river tied together to prevent Soviet naval vessels from traversing the city. Whoever was organizing the resistance knew how the forces might try to control the city. We arrived at the steps of the Duma building to find dozens of T72 tanks parked on the steps, with soldiers still manning the tanks. Their heads were invariably sticking out of the turrets, and their expressions showed more favor to the crowd than animosity. We saw several tanks with bouquets stuffed in their barrels and portraits of loved ones lost in the Afghanistan war lovingly placed on the tanks, along with various icons and other religious symbols. In the old Soviet Union, the public display of religion was widely banned, and very few ever broke that rule. Perhaps the most stunning display was the abundance of Russian flags. There must have been hundreds of them draping the tanks and being paraded by protestors.

Moscow a few days after the attempted coup d'etat. Tanks were parked on the steps of many government buildings including the Duma. Image Credit Jim Cantrell

After a few days in Moscow, we took the train back to Riga. By then, the coup leaders had disbanded, Gorbachev was off house arrest at his dacha in Crimea, and the Soviet anthem was no longer playing on the radio. Boris Yeltsin became the new leader, and the putschists were arrested. On our way back to Riga, the train was full of people returning to their dachas and bringing a full complement of farm animals. Seeing chickens in cages or other farm animals on the train was not unusual. On this trip, however, we were confronted by a full-sized pet pig accompanied by its owner on the train. The pig was very well-behaved, but the owner was not. This very drunk individual, upon hearing our accents and us speaking French, decided that we were a good mark for some money. He tried to convince us that his pig was the smartest swine in the history of Russia and that he could perform magic tricks. We had to pay the owner a small concession fee before the pig would perform. We declined after numerous attempts at convincing us otherwise, but we left that experience knowing that capitalism, at least in this form, was alive and well in the new Russia.

Soviet tanks parked on the steps of a government building near the Soviet Parliament building. Image Credit Jim Cantrell

The Long Goodbye

The trip home to France was unexciting. We had an Aeroflot flight from Riga to Stockholm and arrived early at the airport. Traveling in the former Soviet Union was always an unpredictable adventure; giving yourself as much time as possible was wise. This included going through passport control on the way out of the country. I have always said that you can tell the nature of the country you are visiting by the directions the guns are pointed at the border. Leaving the Soviet Union was always more involved and had more checkpoints and questions than when entering. The message was clear. In the communist days, the ruling elite did not want you to leave the worker's paradise of the Soviet Union. To my surprise, and in conflict with our understanding that Latvia was now an independent country, we were greeted by Soviet passport control officers as we made our way through the airport exit process. They dutifully checked to see if we had the correct visas for entry, examined every page of the passport for evidence of visits to forbidden countries, and after a stern visual examination, stamped our passports, permitting our exit.

After we left the Soviet Visa checkpoint, we proceeded to the airline gate. We didn't get very far before encountering a Visa checkpoint for the newly created country of Latvia. It was a small card table staffed by a young blonde woman. She sat behind the table on a folding chair and had attached a small piece of paper to the end of the table with the word "Latvia" printed on it. As we approached the table, this lovely young lady asked us for our passports. When she examined my passport, she looked up at me and said, "You don't have a Latvian Visa." I explained to her that Latvia did not exist diplomatically when we entered the country. Therefore, we could not have had a Latvian Visa. Without a smile, she offered me a Latvian exit Visa for $200. In those days, nobody in the Soviet Union accepted credit cards, and we always came prepared with lots of spare cash. We each paid the $200 and received a stamp on our passport before proceeding to the plane.

The flight to Stockholm was short on Aeroflot, and this time I was never happier to get off a plane and step into a Western country.

The rest of the trip back to France was long, and when we landed late that evening, somehow, the world seemed completely different. From today's perspective, this impression was correct. The world was now more different than we would ever imagine.

The dissolution of the Soviet Union shortened my career working on the Mars Balloon in France. Late that year, we made our last trip back to Latvia, and we could see that the fall of the Soviet Union was having dramatic effects on our colleagues. We heard from them that all salary payments from the state had ceased and that they could not feed their families except for the money they had saved between their mattresses. Nonetheless, we continued our work with them and pretended as if this would not affect our mission to Mars. Despite this, the country of Latvia was transforming quickly after just a few months, and capitalism was taking root. We made a shopping expedition one evening into downtown Riga and carried around bags of old Soviet Rubles which were deflating by the minute. The largest nomination was a ten ruble note, and many items were thousands of rubles. The currency devalued along with the economy and crumbled before our very eyes.

The real story happening was in the aftermath of the Soviet dissolution and the forces that it unleashed. Social and economic forces changed the world's geopolitical landscapes as the old order gave way to something few recognized, and few foresaw. One of the significant forces I ended up contending with was the lack of pay for my colleagues in the former Soviet Union. The state was not paying my Russian and Latvian colleagues but instead was surviving off the money they made from cooperation with Western countries like France. The French, smartly, had taken to paying the travel expenses of Russians to France during the Soviet days when rubles were not officially convertible to Western currency. There had always been a black market for conversion at attractive rates for Western currencies, but officially, this 'did not exist.' The result was that Russians who came to France were given fixed travel expenses in French Francs, and they would go home with the equivalent of several years' salary for each weeklong trip to France. We understood that this was propping up the Mars 94 project and that it was also a humanitarian gesture to our colleagues. While there was little, we could do to help the country as it grappled with serious economic issues, our small world benefitted from this small cooperation in space. We hoped this would be a seed from which other cooperation with the west on a much broader economic scale could occur later. These were the early days of east-west cooperation, and our optimism would not be well-founded.

My work at CNES continued into 1992, but I began feeling restless early that year and eager to shift from a dying program to something with more promise. My family returned to the U.S. in March, leaving me alone in France to continue my work on the balloon. I knew that the Mars 94 program would not last much longer, and I finally announced that I would leave in May of that year to return to the U.S. The only problem with that idea was that I needed a job to return to. With the dissolution of the Soviet Union, so went the Cold War and the millions of aerospace jobs it supported in the US.

Over the past six months, I have applied to several places I was sure would be interested in hiring someone with experience. I tried ESA, NASA, CSA, and many U.S. Aerospace companies. One company that caught my interest was Martin Marietta. A high-profile employee of Martin was Dr. Bob Zubrin, who had written on the idea he called "Mars Direct." His theory was that by setting a craft on its way to Mars ahead of the crew with the return vehicle, it could remain on the Martian surface generating the return fuel from the Martian atmosphere, thereby saving massive amounts of costs and sending that fuel to Mars only to return. Bob's Mars Direct Concept made sense but was considered by most aerospace insiders as little more than a pipe dream. I found the thinking behind it original and compelling enough to contact Bob.

I traveled to Denver to see Bob, and he arranged interviews for me there. We met in the beautiful Deer Canyon facility, built among majestic red rock cliffs with eagles flying in the air. Since they did more than just NASA work there, the security was exceptionally tight, and I was treated as the equivalent of a Soviet spy visiting the facility. Bob did what he could to improve the situation, but I was tailed and escorted everywhere in the plant, even into the bathrooms. I met several historical characters that Bob had arranged, like Dr. Ben

Clark, who had been one of the original Viking scientists. While the visit resulted in no job offer, I made some meaningful connections here that would again return later to alter my life's course. Having struck out with several other job possibilities, I called my old boss, Frank Redd, one afternoon and asked him for a job back at the space lab at USU. To my surprise, he was very positive and hired me back over the phone. We agreed on a start date and salary, and I set my departure from France in motion.

Starting Over

The one thing that I had not anticipated going back to Logan and USU was the sense of starting life all over again. I had made a comfortable career in France, had friends there, and was very comfortable living in the country. Arriving in Utah directly from France was a cultural shock and a complete change of physical and social living conditions. The relevance of my French and Soviet experience seemed unimportant to most people I talked to as the Cold War was over, and the USA was ready to spend that Cold War dividend on a whole new world where the U.S. didn't have to concern itself with another superpower. Some considered me a 'traitor' for having worked for the French and especially for working with the Soviets in space. Even though the Soviet Union no longer existed, and the Russians were now our 'friends,' my active participation with them made for many suspicions about me and where my loyalties might lie. Fortunately for me, Frank Redd didn't see things that way and instead saw my experience as an asset.

Unknown to me, when I returned to Logan, USU and its space lab (now named Space Dynamics Laboratory) had a small program with the U.S. DoD to initiate space cooperation with the former Soviet Union. I became aware of this one late summer afternoon when a knock on my office door announced Frank Redd's entry. He had a well-dressed and carefully coiffed man about his same age alongside him. Frank, in his typical military-style, immediately dispensed with the pleasantries and introduced his friend with whom he had gone to West Point military academy. As he was introduced,

Don was with the Defense Intelligence Agency and wanted to talk to me about my experiences with the Soviet Union. I was utterly terrified by what was unfolding before my eyes as my paranoia gene was telling me to flee and had me imagining that I had broken some sort of law of which I was not aware.

As I explained very carefully my work in France and the fact that it involved Russians, I could see Don listening with deep engagement, and my senses told me that this was a topic of interest and not the intent to ensnare me in some make-believe crime. Once I gave Don the Reader's Digest version of my career with the Soviets, I paused and allowed him to take over the conversation. Sometimes it's far better to listen than to talk, and I was done talking. To my surprise, Don began by explaining that the U.S. was becoming concerned about the 'brain drain' of Soviet weapons scientists and engineers to places like North Korea and Iran as the Russian citizens were not being paid. I was aware of this predicament, and what Don was telling me rang true from my experiences. Don was part of a program where the U.S. government wanted to put money into Russia and the former Soviet states to stop this brain drain. He was in my office because they needed help from people like me who knew the Soviets. After listening to Don for a few minutes, I indicated my willingness to help. Don and I shook hands, and Frank Redd seemed very satisfied and perhaps even proud of his former student becoming part of some national security program. However, I was not nearly as sure, but it all sounded like another big adventure.

Don disappeared into the ether that afternoon, never to see me again. As a result of this conversation, and only after several short months back in the U.S., I was recruited into a program funded by U.S. DoD and intelligence community to place money into the former Soviet Union to stem the brain drain and acquire dangerous items that may be on the market. We were to engage the Russians through cooperative projects at a university-to-University level, and this would serve as a pathfinder for much larger and more ambitious programs to follow.

The five-year period that followed was my life's most eventful and entertaining. In the early days of this effort, we were some of the first Westerners to see the decaying Soviet weapons and aerospace complex, visited many closed cities, and were given a glimpse of rare hardware such as the Soviet fleet of space shuttles and the lunar landing hardware used in the aborted lunar landing program of the 1960s. In many ways, we were the early predecessors of Space Venture Capitalists heading into the former Soviet Union to buy weapons for sale on the open market, keep them out of the wrong hands, hire anyone with expertise for the same reasons, and employ them on peaceful space projects. Eventually, the work I was involved with and the experience I gained led many internet entrepreneurs to my doorstep ten years later as the New Space Race dawned at the beginning of the 21st century.

SDL already had an enormous task order contract open with what was then known as the Strategic Defense Initiative Office, which would lead this effort in Russia. We had several initiatives underway for what was then called 'Star Wars' technology involving missile defense, so it seemed the perfect vehicle to pass money for the Russian initiative. One Russian initiative was already underway known as RAMOS, which stood for the Russian American Observation System, which sought to understand and replenish the Russian early warning system. As it turned out, the Soviet ballistic missile early warning infrastructure had been allowed to decay to the point where the Russians were without early warning capability for several hours a day. Without money going into the supply chain of infrared systems in Russia, the supply chain disappeared, and satellites that failed were not replaced. Meanwhile, Russia still possessed more than 10,000 nuclear warheads and thus resembled a dangerous but blind wild animal unable to know with any certainty if it was being attacked. It was clearly in the U.S.'s best interest to ensure the Russians could see that we were NOT launching missiles at them and avoid a pre-emptive counterstrike. This whole topic touched on my childhood horrors, imagining that someday the U.S. might be hit with a nuclear barrage from the Soviets. I found this whole topic of working with the Russians intrinsically motivating and fascinating.

Carl Howlett was one of the more interesting people I worked with at SDL. Carl was the father of a college friend and a very talented Program Manager at SDL. He liked to get things done and had a very unconventional way about him that encouraged breaking the rules today and asking forgiveness later. I resonated with Carl. He and I were tasked with producing a Pathfinder satellite program to engage with a Russian University and to engage with SDIO on this to meet the overall national security objectives of the effort. After some brainstorming and research, Carl and I realized that the Moscow Aviation Institute (MAI) had a team coming to the annual Small Satellite conference and would be bringing some hardware for display. We arranged to meet with them while they were in Logan and discuss possible cooperation.

Dr. Gary Popov, MAI's Director of space activities, led the Russian team. Gary is an affable and friendly character especially compared to my earlier experiences with Kremnev in the early days of the Mars 94 project. He had several engineers and professors with him, but Gary was the clear leader. Carl also brought in a new character from our side from the University VP of Research's office. Boyd was a former CIA agent and returned to Logan after he retired from the agency. His claim to fame in the day was discovering the technical bugging operations on the U.S. embassy in Moscow in the 1980s, and he was eager to get involved with Russian programs again. While I appreciated his knowledge of Russian counterintelligence operations, it was clear from the beginning that Boyd had little personal contact with the Russians themselves nor understood how to work with them. I was the only one that understood that and saw myself as an unofficial Ambassador between the two cultures, at least for this effort.

Our meeting with MAI went well and resulted in several technical studies and spacecraft missions to be considered. The Russian counterintelligence was better than we thought then, as the Russian delegation came wholly prepared with launch options and ideas that we could work on together. They somehow had inside knowledge of our intentions. As it turned out, before this effort, Carl and I had been working on using Russian-origin electric thrusters licensed to Loral to fly an SDIO mission to the Moon for Dr. Len Caveny of SDIO. Len had worked with Gary Popov and the company in St Petersburg that initially made the thrusters to bring them to the U.S. and on U.S. Geostationary satellites at Loral.

Len had also worked to import a Soviet-era nuclear reactor to the U.S. for testing, but the real reason was to take it off the market. The good news was that Len already had several ideas in mind and intended to fund these activities. After meeting with the Russians, Carl and I had several phone calls with Len at SDIO and some of his support staff and agreed to move forward. Len wanted to fund many technologies, including space tethers, electric propulsion, and missile defense experiments involving high-speed re-entry.

Our first trip to Russia was in the early autumn of 1992. It was to become an exhilarating but complicated trip as the word had gotten out in the aerospace community in Moscow: Americans were coming with cash to spend. To the credit of MAI, they had put the word out and asked several companies, institutes, and government groups to put together proposals for us to evaluate when we were in Moscow. Carl, Boyd, and I headed over for a two-week trip. For me, it was already a familiar place as it was for Boyd. Carl was in a completely foreign place and was ill-prepared for the sheer onslaught of dinner celebrations, constant vodka drinking, and non-stop meetings. He also needed to prepare for the lack of toilet paper and the need for the anti-diarrhea drug Imodium. Fortunately, I had thought ahead to pack extra supplies for my compadres, and I had pre-exposure to the Russian bacterial environment. More importantly, my immune system was more well developed from all the prior travel there and my years living in France.

For the trip, MAI provided us with a car, driver, and interpreter. Our interpreter was named Olga, and despite the ugliness of the name, it was no match for her actual beauty. Olga was a twenty-something woman who appeared exceptionally well off and well cared for. Her English was impeccable to the point of having a minimal accent. The team felt very comfortable around Olga, except for me. I had enough prior exposure to the Russian Way to be suspicious of her perfect English and sheer physical beauty. Before our trip to Russia, we received a counterintelligence briefing from our security office, and one of the things that they warned us about was women who met Olga's description. They were usually planted by the domestic intelligence agency, the FSB (formerly KGB), and could be identified by their unexplained affluence and exceptional language abilities.

Apart from that, there was no way to know whether Olga reported to the FSB, but we had to assume so. She accompanied us everywhere in Moscow. She went with us to meetings. She went shopping with us and she even went with us to eat in the evening. While I was not highly suspicious of her targeting me personally, I tried to keep more sensitive conversations between the three Americans away from her earshot and only while we were walking outside. Despite his CIA background and what I assumed would be vast counterintelligence knowledge, Boyd seemed to fall in love with Olga. Carl and I spoke privately about this at one point but decided to leave it alone as being 'not our business.' What we were doing was not classified, so we decided that Olga posed no real threat. We would come to regret that decision.

Our interactions in Moscow during this first trip combined individual meetings at MAI with outside groups, excursions to nearby cities to visit plants, and long plane trips to more distant closed cities and plants where they were eager to engage with us. We saw some amazing things in those few weeks. One trip took us to the Soviet spy satellite manufacturer, NPO Kometa, in the Moscow region and to a facility officially off-limits to Western citizens. We also visited an institute making space nuclear power systems. We also visited TsNIIMash which served as an institution that led the Soviet version of the US Star Wars programs. We visited most of the rocket and satellite manufacturers in Moscow. We also met with a US expatriate named Mike Hritsik. Mike was a retired two-star US air force general and had led the treaty verification activities in Russia for the START treaties where both sides agreed to destroy a limited number of ballistic missiles. Mike, born to Ukrainian immigrants to the US, knew his way around the country, spoke the language fluently, and established a capitalist beachhead in the city to work with groups like ours wanting to work in the former Soviet Union. From where we stood then, the combination of the skilled Russian labor force and the meager costs (100x lower than Western standards) meant that the Russian Aerospace market would be hot for at least the next 20-30 years.

Russia and Moscow during these times began to resemble the wild west. It seemed as if everything was for sale, and anything was possible. This sense extended from the streets where Soviet military and space-age artifacts were for sale to large aerospace companies suddenly unveiling what had once been classified information on space and launch capabilities and was willing to sell at bargain-basement prices. It was, in every sense of the word, a Grand Bazaar.

We had two favorite shopping places locally in Moscow: Old Arbat Street and Ismailovski Park. Each place had its usual variety of tourist trinkets like matryoshka, jewelry, and handmade chess sets. If you observed, you could locate stalls selling significant aerospace artifacts like cosmonaut helmets, military uniforms, fighter jet helmets, and spare parts from spacecraft and aircraft. Undoubtedly much of this had been stolen and placed for sale here. Some of these individuals would offer an adventurous tourist the opportunity to see a much more significant private collection for sale if you were brave enough to do it. I once saw an extended collection that involved a long drive in a car outside of Moscow to an abandoned warehouse. There, rows of Soviet-era HIND helicopters stood before me alongside T-72 tanks, various surface-to-air missiles, and small warhead launch vehicles. I was unofficially offered the opportunity to buy Soviet MiG 29 fighter jets and all of the ammunition for Warsaw Pact weapons I could smuggle out of the country. I was assured that getting such devices outside of the country was facile with enough financial support.

We saw much the same thing with our official meetings with aerospace companies and MAI. We were offered 'payload re-entry systems' that looked like they were designed for nuclear weapons, nuclear reactors for use in space, complete flight computers from ICBMs, and complete ICBMs. During one of our early meetings, we even had a serious offer from a Russian firm to sell us six SS-N-18 Submarine Launched Ballistic Missiles alongside the Delta III submarine to launch them. We dutifully reported every offer along with documentation when we could. The period between 1992 and 1996 was a free for all market, and much could be had for an interested and well-monied client. This is precisely what worried the U.S. government.

We made another visit to Russia before we settled on several projects. Boyd came along for the second visit and continued his infatuation with Olga bringing her many gifts and having private dinners. Moscow hotels in the early days were practically nonexistent as tourism in the Soviet Union was not something that many Westerners participated in during the Cold War. Thus, very few Western-style hotels existed. One of the early hotels in Moscow that fit the bill was the Radisson near the train station and Old Arbat Street. There were, to be sure, plenty of Soviet Bloc-style hotels in Moscow but these usually didn't meet Western standards. On this occasion, we could not get rooms in the Radisson and instead were directed by MAI to the Ministry of Defense hotel in Moscow. The price was right at the exchange-adjusted rate of thirty-five dollars per night, but I was wholly unprepared for our experience there. Much like the hotels I had experienced in Latvia, this place was no better on accommodations. A single bed with a radio and occasionally a TV in the room was all one could expect. If you were thirsty, there were no glasses in the room to drink tap water with. Instead, a hallway water dispenser was available with a communal water glass. Breakfast was also a communal affair, served from 0700 to 0900 every morning. Nothing was fancy for breakfast, but it was usually a hearty meal like soup, sausage, and eggs to help defend against the severe Moscow winter. However, the other attendees in the breakfast hall were most memorable. On our first morning there, I counted at least five non-Russian military uniforms, including the Iranian Revolutionary Guard, The Peoples Republic of North Korea, The Republic of Syria, Pakistan, and Cuba. I felt like Colonel Jessup in the movie 'A Few Good Men' talking about eating breakfast a few meters away from soldiers that wanted to kill me. While nothing happened during our week of breakfasts at the MoD hotel, I was sure we had been spotted as American or Canadian and were being watched carefully.

Our second visit to MAI and Moscow ended with the mandatory grand soiree with all the teams attending. It was classic Russian style, with a long table covered with finger foods, plates of unidentified meats and pickled vegetables, and vodka. Lots of vodka. Until this point in my life, I was not much of a vodka drinker. I was not much of a drinker at all. I learned, however, that the Russian culture was built around drinking, and the social event it created was where trust bonds were formed and business deals were consummated. To not partake with our colleagues would render you untrustworthy. In business, ritual drinking was as crucial to doing business with the Russians as the game of golf was to Americans. As it turned out, I was not very good at either. I learned how to drink vodka from the Russians, and in this sense, I was fortunate to learn from professionals. They were also very discerning about their vodka. I learned that there were many kinds of vodka, and the quality varied dramatically depending on the producer. The Russian love for vodka reminded me of the French love for wine. As it turned out, that I managed to appreciate vodka and was willing to drink with the Russians was a very good thing as the rest of my team from Utah State University were Mormons who abstain from drinking due to their particular beliefs. I was the token drinker being the only Catholic in the group.

We returned home from our second trip to Moscow with many ideas. Our Program Manager on the government side was Len Caveny. Len was the Director of Innovative Science and Technology for the Strategic Defense Initiative Organization (SDIO). SDIO was also known as 'Star Wars', the same group just a few years ago set out to build a ubiquitous defense against Soviet missiles and warheads. Now they were engaging, through us, directly with Russians to defend against the chaos that ensued after the collapse of the Soviet Union and the associated brain drain to North Korea and Iran. If the truth be told, the Cold War never ended but instead just morphed into another form of a long Cold War. Len was the perfect leader for this effort at the perfect time. He had prior experiences with Russia in the electric propulsion arena and with Gary Popov at MAI. Len was a former naval officer, and by the time I met him at the later stages of his career, he had developed a wicked sense of humor and a desire to truly bury the hatchet with the Russians after the Cold War. In his career, Len had worked propulsion, launch systems, and general technology surrounding our ability to locate and shoot down incoming warheads. Len was a tall and thin man with a taste of red hair seeping out of an emerging field of gray but always with a strong sense of humor. More importantly, he was a brilliant engineer who got things done. Given his air of scrappiness, one could easily imagine Len being very experienced in bare-knuckle fistfights on ships and in ports. His role at SDIO was to be the 'bleeding edge' of programs and to develop this new relationship with our former enemy.

One of the projects that Len was involved in when we first met was a possible purchase of mobile nuclear reactors from the Soviet Union known as Topaz II. A model of this reactor was initially shown at a Department of Energy Symposium in Albuquerque, New Mexico, in January of 1991. Len and his SDIO management later agreed to purchase two of these reactors for use in an in-space electric propulsion demonstration. The idea behind purchasing the reactors was essentially to keep them off the world market and learn what we could from them. This was the broader theme of our work in the former Soviet Union. American bureaucrats got involved once the agreement was made to make the deal come apart. First, the model displayed at the DoE symposium could not be exported back to its country of origin, which seemed to defy all trains of logic. Second, concerns over using nuclear reactors in space began to scuttle the purchase of the two Topaz II reactors. Secretary of State James Baker eventually got involved, and the sale never went through. Len never let things like this deter him. Rather, it energized the man to make cooperative space programs between the two countries happen.

The time between the second and third visits to Moscow was a time of infinite possibilities and big ideas. As a result of our visits, we had several projects that we were to consider. First, there was the idea of a re-entering spacecraft that would emulate a re-entering nuclear warhead and measure its ultraviolet signature from within the satellite. This well-known phenomenon was known as 'UV bow shock' in the U.S., where the high-speed projectile encountering rarified atmospheric molecules would heat them violently and split them apart so that ultraviolet light was emitted from the bow wave as the object passed through the atmosphere. SDL had done two such experiments at lower altitudes and speeds, and this experiment would extend the altitude to 500 miles and speeds of 25,000 miles per hour. This interested the missile defense crowd as the Earth's ozone layer blocked all UV light, and a nuclear warhead's bow shock could be seen from above against a very dark Earth target. This technique was made for detecting nuclear warheads mid-course, even if they had been cooled to hide their infrared signature. The second project involved flying an electric propulsion-powered satellite to the Moon, and the third project involved creating 'Earth Return' capsules coming back from the Mir Space Station. As the station was called, Mir translated to the Russian word for 'Peace' and was put up in the Cold War but still flew with Russian astronauts. The U.S. had a space station called Skylab back in the 1970s, but due to the shuttle being late in taking to the skies, the station re-entered before it could be boosted up in altitude. There was a great deal of interest in both of these programs, but our mission was to stem the brain drain from the defense sectors more than the civil space side, so we engaged mainly on the UV bow shock experiment.

Instead of calling the new experiment something unimaginative with an indecipherable acronym, we decided to call the new experiment Skipper. It was named as such because the spacecraft was designed to gradually 'Skip' over the atmosphere from orbit, gradually probing deeper and deeper into the upper reaches of the gases surrounding our planet. Each time the spacecraft passes through the atmosphere, a lot of friction from aerodynamic heating builds up. This heat causes the surrounding atmosphere to glow in the ultraviolet spectrum.

Our third visit with MAI was intense, and a long week of decisions to be made. It was also interrupted by what I came to call 'Lenin's Revenge': dysentery. Most Americans cannot travel to far-flung parts of the world without experiencing intestinal distress. Our upbringing in North America seems to limit our exposure to various types of germs, viruses, and spores. As a result, our immune systems are highly specialized for the given family of environmental effects we encounter in North America. Go to Mexico, and we get 'Montezuma's Revenge.' Go to Europe, and we seem to be fine. Venture further east, and it's once again like going to Mexico. I had spent several years prior in France and had been exposed to a wider variety of germs. As a result, my immune system was far more robust than my colleagues, and when we would go to Russia, I would not get sick as often as the rest of them. I cite this experience even today as one of the reasons I don't get sick very often. I made this case in 1991 in Toulouse, France, when I was invited to the U.S. Chamber of Commerce's 'Wine Tasting and Food Pairing' event in downtown Toulouse. At this event, the local business glitterati and political establishment attended to spend the evening tasting wines paired with various cheeses—French cheeses. The local "Fromagier," or Cheese Master, was the event's emcee. One of his speeches amidst the wine tasting was railing against the then-European Union's pending rules restricting all EU cheese to be pasteurized as Mssr. Fromagier explained this would essentially kill off half the French cheese industry as many kinds of cheese required bacteria for the taste, and pasteurizing them would make them taste like 'le merde' (translated as 'shit'). I found this man's rebellious spirit compelling and enjoyed listening to him. As the evening wore on and I became consumed by more wine. I began talking to him about my experiences in Russia and how my U.S. colleagues always got sick, and I didn't. Perhaps it was due to the exposure to bacteria in French cheese and the generally less sterile culture in Europe. He agreed and, to my surprise, put me up in front of the 300 or so people attending the event to make a speech about my experiences … in French. In my drunken state and speaking patois French, I explained my theory. To my surprise, I received a warm standing ovation! I had hit a nerve with the French and was the sole celebrated American in attendance that night.

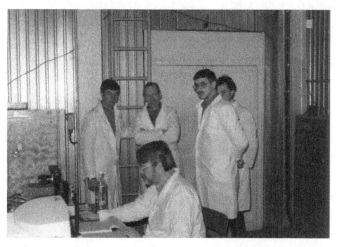

Skipper test team at Baikanour. Image Credit Jim Cantrell

Meanwhile, back in Moscow, I led meetings with the Russians in the beginning days of our two-week-long visits. My colleagues were busy being sick and hugging toilets while I fended off dozens of Russians, or at least almost by myself, as Mr. Kay Jeppeson was with us. Kay was the CFO of our space lab and had come along on this trip to negotiate out of curiosity. Kay was a quiet man and played the role of a quiet accountant very well. He was also in unfamiliar water and hung close to me during the discussions and breaks. We had made a few visits to Moscow now, and it was time to come home with a deal. There was much pressure from Len Caveny back at SDIO to get a deal in place, and we were expected to call him every evening with updates. So, Kay and I sat at the negotiating table with a team of seven Russians, including Gary Rosovsky, our translators, and the NPO Lavotchkin team that would build the Russian contribution to Skipper. As it would turn out, these were the same Russians I had worked with on the Mars 94 program back in France. I had not sought them out when we were looking for projects to fund in Russia, but when they did show up, and I was reminded of their complete professionalism and deep experience in space systems, I warmed up to them. Seeing familiar faces amid a new and strange reality was good.

Kay and I hammered a deal over the four days our colleagues missed. We started by discussing what we wanted to be done and then proceeded to the part about how much money we would send to MAI and the rest of the team. Kay was a master negotiator, and I let him handle the money negotiations while I negotiated the technical scope. We knew how much Len had to spend, and the Russians thought they understood Western pricing. It was an exciting game to engage in. We started around 10 AM every morning after we arrived from the hotel to the office. We had the mandatory coffee and cookies before we started and eventually would engage in business discussions for about an hour before lunchtime. We walked over to the MAI cafeteria in the snow and sat at a long table for a Russian-style lunch. The food was always good with that 'ah' smell that greets you on a cold winter day. However, lunch was never speedy, and we rarely completed our meal before 2 PM. We would walk back to the office, engage again in the mandatory coffee and sidebar discussions, and finally sit down to negotiate by 3 PM. For us Americans not used to this social style of negotiations, it was making us crazy. I sometimes wondered if this deliberately broke down our will to resist continual demands and new conditions. I could see the fatigue in Kay's eyes every afternoon, and he could see me getting sleepy around three every afternoon when maximum jet lag sets in. However, we soldiered on and started to make progress after a few days.

On the fourth day, Carl gathered enough physical strength to come to the meetings and help us negotiate. Kay and I had taken it as far as we could go. For the Russian contributions of what amounted to launch, a considerable hydrazine propulsion system, solar panels, and battery system, we were down to 3 million dollars. We understood that the Russians had no cost basis for anything they were asking; for them, it was pure sport. On our side, on the other hand, we had a budget to make, and Len had authorized 5 million dollars for the entire project. We knew that the SDL side of the effort would cost us 2.9 million dollars, so we had to find 0.9 million dollars to cut. Kay and I briefed Carl on where we managed to get the MAI price down and the remaining gap. We developed a quick strategy to close the gap. Kay and I would scrub the SDL numbers to find $500K. Carl would take a fresh swipe at MAI and try to get concessions on the order of $500K. That would put us in range of the budget that Len had given us, and we could go home and start building.

Kay and I reported back after lunch, having found the necessary reductions, but Carl had yet to find a way. MAI was holding fast to their pricing and adding more demands. We sat and argued with MAI for several hours, and finally, at about 6 PM, we took a break. The three of us went for a short walk outside to get some fresh air and discuss our strategy. When we returned, we arrived to find the negotiation table decorated with several bottles of whiskey and unique shot glasses. This was meant to impress us as the 'regular' vodka was dispensed with in favor of rare and pricey bourbon from the U.S. I knew this meant it would be a long evening. We were also greeted by what now apparently was heretofore a silent partner in the negotiations: a member of the Soviet Academy of Sciences. Unbeknownst to us, the good Academician had a finger in the approval process of the mission and had some financial objectives of his own. Back in this era, you could always tell who the powerful were in the Soviet Union by the nature of their dental work. If they had visible gold crowns, they were a part of the state security, the politburo, or the extended power structure like the Soviet Academy of Science. In this case, our good Academician had a mouth full of gold and was either KGB or otherwise very influential. Seeing his smile, I knew we had a long way to go.

The evening continued, complete with the whiskey drinking, toasts to the greatness of the Russian and American societies, and the pleadings from our MAI colleagues led to an impasse. We had no choice but to shut our briefcases and head home. We had plane fares the following day out of Sheremetyevo and opted to leave the negotiations by about 10 PM. Heading out in the late-night Moscow snow and back to our depressing hotels was one of the lower points of my career in the former Soviet Union. When we returned home, we had a long meeting with Len Caveny and discussed the situation. He encouraged us to negotiate at a distance and conclude before we returned with a return contingent on a contract ready to sign. We took Len's advice and began that process the following week.

The following week went fast, with plenty of things to do and personal life that had yet to take time off while I was away halfway across the world. My kids were always happy to see me when I returned and loved the various gifts I would buy when I was there—for oldest daughter liked the handmade dolls that you could buy at Ismalovsky Park north of Moscow. She also loved the Russian stacking dolls. My son liked the car toys and the board games. Coming home, I usually found a home that was a complete mess, broken appliances ready for my attention and social obligations that I had not planned for or wanted. I tried to take time off after such trips, but in the end, it wasn't easy, and I managed to work nearly seven days a week despite my efforts to avoid this.

On the Sunday after our return from Moscow, I went to work to retrieve some documents to work at home. Typically, SDL was a tranquil and sleepy place on weekends, especially Sundays. Imagine my surprise as I rounded the hallway and looked up to see Olga our translator from MAI. I surprised her almost as much as she surprised me. I asked her what she was doing here in the U.S. and specifically in the hallway on a Sunday. She seemed to be at a loss for words and struggling to get an explanation when Dan, our IT Administrator, walked around the corner and greeted me with his big smile. If God ever made someone deliberately naive, this would be Dan, and I always thought that this made strange bedfellows with him overseeing IT. However, in the 1990s, IT was more of a technical field than a security profession, so it fit the era. In Dan's presence, Olga explained that Boyd had brought her to the U.S. to 'learn about computer networking' and that Dan was kind enough to teach her what she needed to know on the SDL network. My Spidey senses went off due to this incident, but I decided that perhaps I was too paranoid about all of that KGB stuff I had learned over the past 20 years. I welcomed her to the U.S. and thanked Dan for helping our program.

Work resumed the next day as it always did on a Monday, with program meetings and many emails to answer. This regular cycle ended late in the morning with a knock on my office door. Carl Howlett was at the door with what turned out to be a pair of FBI agents. Carl introduced me to the agents, who wasted no time showing me their badges as if they were the one thing, they were most proud of. They came into my office and sat down at my desk. The FBI agents asked me a few questions but quickly realized I was unaware of the 'situation' that brought them to SDL. As a witness, I was not very useful to them, and they quickly left. However, I was able to piece together Sunday's events from Carl and others. Much had happened the day before after seeing Olga in the hallway with Dan.

On Sunday afternoon, Boyd arranged to collect Olga from the University offices where Olga worked with Dan. He then took her to the off-campus offices where classified data on several satellite programs were stored. For reasons I don't understand to this day, Boyd decided to take Olga on a tour of the classified data storage facilities, which were classified as Top Secret. Along with such a classification level goes a certain level of alarms and safeguards. Olga and Boyd engaged in sexual relations inside the classified storage facility. The shrieks of joy emanating from Olga set off the glass break sensors and triggered a silent alarm. This alarm required the police to respond with weapons drawn within 10 minutes, which was short enough time to catch the two love birds in the carnal act on the data center floor. To make matters worse, Boyd's wife had become suspicious of his travels and communications with Olga before this event. She had made an adverse information report to the CIA, his former employer. The FBI was already on the case when Boyd and Olga decided to tempt fate. The remainder of the James Bond saga remained very quiet, and I noticed that he was not seen anymore after this event, and neither was Olga on future trips to Russia. I assume neither of these individuals fared well due to this incident and were terminated.

Our negotiations with the Russians continued over the next few weeks, with the MAI team finally agreeing to a price we could accept. We were notified by a fax transmission that MAI accepted the price and was ready to sign a contract. Carl, Kay, and I returned to Moscow the following week and completed the contract. While much water had gone under the bridge by that point, it was a significant relief to have this part of our relationship solidified and technical work could begin. We made a short trip to this episode and quietly signed the contract in the same office where, weeks before, we had drunk bourbon until the wee hours of the morning.

Skipper started with a large meeting in the U.S. It was early spring in Logan, Utah, and the Russian delegation came to the meeting in full force with twelve people. We were still a small team on the U.S. side with four engineers and Carl, our Program Manager. We did have many U.S. government representatives in classic style that out-numbered our U.S. design team, so the match was more than adequate. The Russian style in these meetings was something that we needed to become used to. They were very aggressive at a group level and confrontational individually. I did not know most of the Russian delegation, and it seemed they had a Cold War chip on their shoulders. I would come home from a day of meetings with them and have minimal left to give to my kids or energy to do anything other than sleep. Our U.S. delegation was equally aggressive but headed by Len Caveny, who made things smoother. His right-hand man Doug Allen, a friendly but cautious man, came from W.J. Schaeffer & Associates and was a calming influence on the group. We also had our share of 'state department' employees that would attend the meetings with the oft-stated objective of 'making sure that

U.S. Foreign Policy is followed.' These folks we presumed were from one of the many U.S. intelligence agencies, and Len referred to them as 'the folks upstairs.'

It quickly became apparent that our working meeting with the Russians was filled with a serious amount of misjudgment and trust issues. Linda Allen (no relation to Doug) was our avionics lead engineer and offered to host a party at her home for the team. Our first party at her house had a large contingent of Russians, the U.S. team, and many other Americans from work. As predicted, the booze flowed, lots of food was consumed, and relationships between the Russian and American teams began to grow. One of the attendees was a mutual friend, Frank Walker. Frank was a former submariner who seemed to face life with a smile and a sunny-side-up spirit. Always the joker and kid in an adult body, Frank brought some surplus helium weather balloons and various things we could launch on them. This was very popular with the Russians, who had drunk heavily. We started by sending off flashlights which, given the balloon's twisty ascent, appeared more like a UFO than what it was. The Russians then got into the spirit of things and started launching notes on the balloons to long-lost loved ones and, in some cases, denunciations of people they had been bothered by. The party wound down by about two in the morning, and we headed into the weekend.

The Russians had planned to stay for two weeks, and only some of us wanted to entertain them all weekend. The solution to this dilemma was to give them the keys to a van from work. They took great delight in driving the van all over the valley, into the mountains, and shopping. The Russians loved to shop but buy nothing. This whole scenario had the potential to end very badly, but this was the early 1990s, and things were somehow more relaxed in this small Northern Utah town. The driver, our interpreter Yuri, was skilled enough not to get into an accident and managed to drive well. He did report being pulled over one evening by a Sheriff for failing to signal a turn. After much discussion, the Sheriff decided to let them go even though not one of them had a valid driver's license. I had to smile and wonder what went through the older officer's mind as he came across a van full of Russians driving a university vehicle with no license. Sometimes it's better to let things like this go than to deal with the drama it might otherwise incur.

The following week of meetings went well, with a better working relationship developing because of booze and social interactions. My management chain of command was all Mormons who abstain from drinking but somehow found it amusing and valuable enough to support our after-hours parties and offer to pick up the tab for the food and booze. Despite the evolving sense of camaraderie with the Russian team, the technical meetings all seemed more like negotiations than designing a satellite. We had a lot to work out. The Russians were to build the propulsion system for the satellite, provide the launch aboard a Molniya rocket from Baikonur Cosmodrome, and the U.S. side would build the satellite bus and its instruments.

Our satellite was designed to launch high above the Earth and then lower the orbit to 'skip' off the atmosphere in multiple small maneuvers. During these skipping maneuvers, we would measure the ultraviolet light emissions of the bow shock that the satellite produced while flying through the atmosphere. Our ultraviolet instruments looked out from the satellite's heat shield through small quartz windows into the 25,000 MPH slipstream of the atmosphere. We would gradually lower the satellite further into the Earth's atmosphere until it finally burned up in a final fiery betrayal of the machine, much like Joan of Arc being burned at the cross. This experiment aimed to simulate an incoming nuclear warhead and obtain valuable phenomenological data that we might later use for systems providing early warnings of nuclear attack. It was a strange change to be sitting with what amounted to former Soviet engineers working on a space system that would warn each country of the other country's attack on it. Times had indeed changed, and at this moment, I realized how different a world we had become.

Design work on Skipper began in earnest in the spring of 1993, ready for a late 1995 launch. That gave us just under three years to complete the design. It was a complicated spacecraft that amounted to a 'flying fuel tank.' Overall, its shape was cylindrical and designed to mock a re-entering warhead with a domed re-entry shield made from aluminum and body-mounted solar panels. It would nominally spin about its cylindrical axis, and the spacecraft would be inherently unstable because it had a long shape. The propellant inside its tank would normally slosh around during its spinning and cause energy loss, destabilizing its spin about the long cylindrical axis. We devised a control system that used a combination of the Earth's magnetic field and small gas thrusters to control the spin axis gently. This was still something akin to a plate spinning on a stick and, without constant attention, would fall off. Much else about the spacecraft was unusual given its unusual mission and, in many ways, resembled more of a sounding rocket payload than a satellite.

Our early efforts to befriend the Russians paid off as time went on. Despite the rocky start with Olga and Boyd, the two teams grew close, and the trust between the people developed slowly over time. We had technical interchange meetings between the two teams roughly every other month, and they alternated between the U.S. and Moscow. I became very familiar with the city and language and could wander around Moscow using public transportation. Sometimes we stayed in hotels, and sometimes, people rented their apartments to us in a post-Soviet version of VRBO. The Russians got used to having their transportation in the U.S. and continued to drive our company van on every trip to the U.S. They became gradually more adventurous with each trip and, on one summer weekend, decided to venture up into Yellowstone National Park on their own.

Having had much experience in that part of the world, I understood the inherent dangers of the wildlife, especially the bears and the buffalo. I explained as best I could that these creatures could and would kill them given the right circumstance. The Russians seemed to wave off any concern as apparently any Russian worth his or her salt is more than aware of the dangers that bears posed. After all, we termed the Soviets 'the Bear' during the Cold War. The Russians didn't have buffalo there, however. I explained to them not to approach buffalo any closer than 100 feet, and they listened politely as I described how calm they can be and then suddenly deadly. I was relieved when they returned to work on Monday following their trip with the entire crew intact. However, I was horrified by a picture my friend Ioury Bojor showed me with the Russian team kneeling just below a rather large male buffalo no more than 5 feet behind them. You could see the steam coming from the buffalo's nostrils and the large smiles on the crew's faces, which is hard to obtain with Russians who are not prone to smiling in public. I didn't bother continuing my earlier lecture but rather shook my head and remembered a phrase my mother often used when she would see the stupid shit I did as a teen: 'God looks after Fools, Drunks, and Babies.' I could only imagine which one of these applied to my colleagues.

Carl did an excellent job bringing the program along for a paltry sum of money for the time - 5 million dollars to pull this entire mission off. While the goal was to have a pathfinder for the U.S. DoD to cooperate with the Russian MoD, we had the technical objectives of a credible flight program and a fixed budget to meet. As with all projects, there are a million things along the way that would 'improve' the product or make it 'better.' Engineers are born like this and know no other way. Usually, a good Program Manager would lecture his team about 'better being the mortal enemy of good enough,' but some, like Carl, are still engineers at heart and fall into this trap. When Carl presented a wish list to Len Caveny that amounted to one million dollars of additional costs, Len said no. Carl was not the kind of man to take no for an answer. He was very stubborn. Carl had started his life as a devoted Mormon but later decided he did not believe and no longer wished to live according to the faith's strict rules of behavior. This led to his ex-communication and fed an already strong non-conformist streak in him.

An example of this was something that became a local news topic when Carl decided he didn't want to wear clothes while mowing his lawn. He had 8-foot-tall fences around his yard, and he reasoned that nobody could reasonably see him doing what he wanted to do in his private backyard. With more time on their hands than common sense, the local neighbors began peeking through holes in the wooden fencing and calling the police to report public indecency. Carl was arrested and went to court and prevailed. The judge ruled in his favor because he had very tall fencing, and one would have to take extraordinary measures to see him mowing the lawn in the nude. Unfortunately for Carl, Len Caveny was a little less understanding when Carl took their private disagreement over Skipper to the trade publication Space News. The end was swift for

Carl, and he was replaced after a two-hour meeting between Len Caveny and the Lab's Director, Alan Steed.

Shortly after Carl was removed as Program Manager, I led a delegation of engineers to Russia to conduct a Technical Interchange Meeting. Our interim Program Manager was Dave Burt, who was an engineer's engineer. Dave was not known for being friendly, but he was a soft man beneath his gruff exterior. Other duties kept Dave at home on this October trip, and I was happy to lead the delegation. The meetings went as planned, but the trip to Moscow was anything but ordinary.

I had taken to staying at the Radisson hotel near the center of Moscow. It was close to many good restaurants, and the feeling of being close to the west in the heart of Russia was comforting. Russia was changing rapidly following the fall of the Soviet Union. Capitalism was beginning to take hold and especially in what was once the black markets. Early on the street, criminals were not much to be feared, but a few years of experience began to breed a new level of criminality. Many former KGB had come into sudden wealth, finding ways to exploit and extort Western businessmen, especially the emerging Russian business class. What became the 'Russian Mafia' eventually ran the streets at night and was known for its violent tactics. Fortunately, we didn't encounter them in the aerospace sector, but we rubbed shoulders with them at nightclubs. I found them sobering and was very careful not to be noticed by them. The other thing that was happening was inevitable: the communists being left behind were becoming angry and vengeful.

Yeltsin was running the country like a freewheeling capitalist, and this angered many of the old-time Soviet hardliners who wished for the former days of repression and Soviet-style calm. This conflict came to a head in the Russian Duma, roughly the equivalent of the U.S. Congress. Soviet Hardliners controlling the Duma faced off with Boris Yeltsin. The Russian constitutional crisis in 1993, also known as the 1993 October Coup, resulted from a power struggle between Russian President Boris Yeltsin and the Russian parliament. The power struggle reached its crisis on 21 September 1993, when President Yeltsin attempted to dissolve the Duma, the country's highest body and the parliament. In response, the parliament declared the president's decision null and void, impeached Yeltsin, and proclaimed Vice President Aleksandr Rutskoy to be acting president. We tried to ignore this political mess, but we knew it threatened what we were doing if the results did not fall in Yeltsin's favor. By the 3rd of October, demonstrators had removed police cordons around the parliament and taken over the mayor's offices. Initially declaring its neutrality, the army surrounded the Duma, also known as the White House, with T-72 tanks. These tanks could be seen on the streets of Moscow.

Our hotel was not far from the Duma, and we were aware of the area's various roadblocks and military activity. It caused us some difficulties getting to and from work as the traffic was closed in this area. On the morning of 4 October, I decided to walk before heading off to the meetings. As I woke and put my walking clothes on, I could hear the distinct sounds of artillery shells firing and landing on some unfortunate target. I was unaware the Russian T-72s were mounting an assault on the White House as I walked out of the Radisson. I could hear the shelling breaking through the chilly October morning and could hear the thuds impacting somewhere. What was most amazing to me was the ordinary Russian citizens who walked to the store to make their daily purchases alongside others heading to work. They were ignoring the tanks and the shelling going on only several thousand feet away. My curiosity got the better of me, and I headed up the walkway along the Moscow River.

As I approached the bridge over the river at Kutozovsky Avenue, to my amazement, there were three to five T-72 tanks with their turrets pointed north at the White House. I was shocked to see and feel the power of one of the tanks fire on the building in the distance, which was now on fire in various places. The shelling was not constant but punctuated by about 15 minutes of quiet. You could hear small arms fire coming from the Duma in directions that were not clear. Again, my curiosity overcame my better judgment, and I followed a number of people who seemed determined to cross the bridge on foot behind the tanks. No police or soldiers were present to prevent our movements. Only fear for your life did that. As I mounted the bridge, many people were spectating and waiting for the next shell to be fired. It was almost a circus-like atmosphere. I stayed up on the bridge long enough to experience up and close the firing of the southernmost T-72. The concussion was deafening while it was quickly 1000 feet away from me. What was more impressive was the feeling of the bridge vibrating underneath my feet. The inner engineer in me was sensible enough to deduce that this was potentially a perilous situation, and I left.

I headed back towards the Radisson and took the enclosed footbridge across the river. I left the bridge past the Turkish Embassy, which was right on the river. I wanted to see the shelling from a safer vantage point, but I was closing in on the White House. The small arms fire was getting louder and sounded much like 50-caliber arms. These are usually powerful enough to punch holes in engine blocks and penetrate light armor. When they hit a human being, the results are horrifying. As I crept closer to the Duma, it became apparent that the shots were being fired in my direction and at a group of soldiers on foot 2000 feet ahead of me. I could hear the rounds striking trees and concrete abutments in the park and finally decided that I had seen enough. I ran back to the hotel with my flight gene in complete panic mode. After returning to my room, I managed to shower and meet my colleagues at 10 AM in front of the hotel. The bus took us around the fighting, and my colleagues were eager to know what was happening. I remained silent, partially out of not wishing to admit how reckless and stupid I had been that morning. We left two days later, the shelling had ended, and the White House was burning. I was thrilled to go on that trip and knew I had somehow experienced a random encounter with history. The ten-day conflict I witnessed became the deadliest single event of street fighting in Moscow's history since the Russian Revolution.

In the U.S., our new PM was announced as Bruce Peterson, and I was promoted to Deputy Program Manager. I was not convinced this was a promotion as I preferred to stay in the engineering side. However, since I often disagreed with my superiors, having more decision authority was good. I also took on a higher profile working with our Russian colleagues. This meant more travel to Russia, and more debriefs on what we saw there. Part of the consequence of working with post-Soviet Russia on space and missile defense joint programs was that the U.S. intelligence agencies were naturally interested in what we learned while in the country. I soon developed a relationship with the National Air and Space Intelligence Center (NASIC), the U.S. Air Force unit, for analyzing military intelligence on foreign air and space forces, weapons, and systems. NASIC assessments of aerospace performance characteristics, capabilities, and vulnerabilities shape national security and defense policies and support weapons treaty negotiations and verification. Post-Soviet Russia greatly interested them, and our on-the-ground insights and observations were unique and valuable. This relationship also added additional stops on my flight itinerary to and from Russia, stopping in Dayton, Ohio.

By mid-1993, Skipper was well on its way to becoming a reality. We were deeply involved in the design, and many critical decisions were being made. Our interchanges with the Russian teams were now almost monthly. The Russians would come to Logan in masse for one month, and the following month, we would go to Russia for a week. I spent more time there than the rest of the team to take care of 'overhead tasks' of managing the relationship with the Russians and keeping the Defense Attaché at the U.S. Embassy apprised of our activities. I quickly felt my involvement becoming less and less technical and more and more bureaucratic. Still, it was all fascinating and led to new experiences I would have never otherwise experienced.

One of those experiences was a Russian 'banya.' The word 'banya' in Russian means 'bath' in English. However, there are more than simple subtleties between the two realities. I discovered this reality one summer weekend when I stayed for an extended set of meetings in Moscow. My boss Bruce and I had opted to remain in Moscow for two weeks to take care of several technical and management issues rather than commit slow suicide by traveling back to the U.S. week after week. Knowing we were remaining for the weekend, my friend Ioury from the project invited me to a 'Russian banya' on Sunday. Not knowing what it was, I accepted and asked if Bruce was coming as well. Ioury's response was both instructive and frightening. He replied that Bruce was uninvited as 'he would not understand this kind of pleasure.' Given that Bruce was Mormon and I wasn't, I assumed this had something to do with alcohol consumption. I was partially correct.

Ioury had instructed me to take the green subway line to the Rechnoy Vokzal northwest of the city. This was beyond the ring road and starting to surpass what Muscovites would consider 'civilization.' Ioury didn't instruct me on what to bring, but I assumed that this was a spa experience, such as a bathing suit, towel, and bourbon. I even brought a hat. I arrived at the station at about 0900 as planned and found Ioury smiling as he greeted me, getting off the metro. We dispensed the pleasantries and headed outside, where we were greeted by three of our other colleagues from Lavotchkin. Outside every metro station during this period, there were small kiosks where you could buy supplies and every manner of things you couldn't find in regular stores. We stopped and bought some Russian Baltika beer as well as Krystal Vodka. Something told me an afternoon of heavy drinking was in order.

After finishing the purchases, we mounted into an old Lada station wagon and headed west. We drove for about an hour and ended up in a very rural area with just small villages and dachas. Being a student of history and strangely fascinated with the Nazi attack on Moscow in 1941, I knew that we were in territory once held by the Wehrmacht. After a lengthy drive where scarcely anyone said a word, we exited the highway and entered a small village. The village had 20-30 tiny wooden homes that Russians would call a dacha, or country home. Usually, these are only occupied rarely during the week but reserved for weekends and vacations. In many ways, these places were out of the eye of the ever-present KGB and state authority and were a place for Russians to be more themselves. It seemed to be a pressure relief valve for the political and societal repression that they suffered and a way to connect with simply being human again.

We stopped at a dacha at the end of the village, and my colleagues explained that it was Alexander's dacha and that his father lived there. I imagined the 'banya' was nearby, but I couldn't spot it. We entered the 100-year-old unpainted wooden cabin, and I noticed that it didn't have the ordinary concrete floors we have in the U.S. but rather wood planks on top of dirt. It was very well kept but reminded me of fishing cabins I had seen in the U.S. Alexander's father greeted us warmly and paid particular attention to me. He explained that I was the first American he had ever met and that he was honored to have me. He also brought out an old rifle that was so corroded I doubted it would fire. It was a U.S. M1 carbine that was part of the arms and supplies the U.S. sent to the Soviet Union during the war. Alexander's father explained to me in very patois Russian that he used this to kill Nazis when he was young and wanted me to hold it. I did, and much banter then took place as we all smiled at each other. I gave the M1 back to the elderly man, and he proposed a toast. Knowing that every Russian social event begins with a toast is essential. This time they found some bourbon and toasted to me and the United States. It was a sincere appreciation of my country's history in fighting tyrants; their words were heartwarming. At this point, I could never have imagined that our two countries had been poised to destroy each other for almost 50 years.

We drank the bourbon first and then the Krystal Vodka. The latter was rumored to be the best Russia could offer and did not disappoint. After about an hour of relaxing and drinking, Ioury said it was time to head to the banya. I grabbed my bag, and we headed out the door and crossed the street into the forest. It was a beautiful warm summer day, and the sun felt good on my exposed skin. The forest was dark as we walked along a narrow path toward this unknown place. My mind started wandering after about 10 minutes on this trail and the reality that the banya was not part of the dacha complex. I trusted Ioury more than most people, and I had to rely on my sense of trust not to begin asking probing questions and revealing my increasing concern. Call it a sixth sense, if you will, but my senses told me this was not what I expected it to be. We passed into an opening in the forest and by what looked like the local graveyard. Indeed, a hole was dug, ready for someone to take an eternal resting place. This did not enhance my sense of well-being at all.

I asked Ioury how much longer it was to the banya; he thought it was another 10 minutes. About 20 minutes later, we came upon an enormous clearing in the forest with an extensive planting of potatoes. There were 5 acres or more of potatoes. We stopped at the edge of the clearing, and Ioury told me in Russian and pointed across the clearing, 'Vot banya.' I could see a small wooden shack with smoke coming from what appeared to be a chimney.

I also saw a middle-aged man dressed only in his underwear, hat, and sandals in the potato field. These were not sporty and attractive boxer shorts but Fruit of The Loom 'tighty whitey' underwear and were offset by a large belly that could only be described as a 'beer belly'. Oleg stopped hoeing the potatoes and began walking in our direction. As he greeted the group, I was introduced, and as he smiled, I noticed that half of his teeth appeared missing. I was starting to get a tad worried about this banya experience, but again I relied on my sense of trust in Ioury. Despite that, this whole experience was getting stranger by the minute.

The group walked over to the shack in the shade of the birch trees. As we reached it, Ioury began to give me a tour of the place. There was a sizeable flowing spring adjacent to the shack, which was used to cool off. Many birch branches were gathering like primitive brooms and hanging upside-down in the shade of the trees. I didn't ask what they were for. As we entered the shack, there was an anteroom with some chairs and benches, and it was generally open to the air. A door inside this room led into Dante's inferno. Inside, the temperatures easily exceeded 130 degrees F and were heated by a wood-burning stove. The air from the doorway was so hot it felt like it was burning my skin right off my body. Ringing the edges of the room were wooden benches covered by wool blankets. I spent only 20 seconds or so in this room and exited. The Russians seemed to enjoy the idea that I could not handle the heat very long, and all got a good laugh out of my shock at the air temperature.

As soon as we arrived, the bottles of Baltika beer went into the small pool formed by the spring, and the water was cold enough to make the beer very palatable. We all grabbed one and drank it quickly. Our walk here had seemingly worked off the buzz from the bourbon and vodka and generated quite a thirst. The first beer went down very quickly. The second one went almost as fast, and the familiar buzz was starting to return. Ioury announced it was time for the banya as we finished the second beer. Being the modest American I was, I went around the corner to change out of my street clothes and swimsuit. As I rounded the corner back to the banya, I was faced by five naked grown men, all drinking beer and talking. Now my flight gene was in full force. I had to think quickly as I sensed a situation beyond what I was prepared for.

Again, I trusted Ioury and believed he knew I was painfully heterosexual. So, I was able to discount that fear quickly. I finally decided this was like my experiences in France and Spain, where the beaches were generally clothing optional. I wasn't too ashamed of my nudity and thus decided that it was best to 'do as Romans do when in Rome'. My swim trunks that I used in the hotel pool came right back into my bag. I grabbed a towel and wrapped myself as we entered the banya.

We started in the anteroom drinking more Baltika. The conversations were mostly in Russian but slow and simple enough for me to understand some topics. Much of it centered around Oleg and his potatoes. This banya belonged to Oleg and his family, and he was proud of it. After our first beer, we all went into the banya and sat our naked behinds on the wool blankets. The wool was intended to insulate our flesh from the wood surfaces and to prevent burns. I could only stand it for about 30 seconds and then headed out. This was followed by a round of laughter and comments about the softness of the American in the group. I didn't care. As I sat down in the spring-fed pool, I could almost hear the water sizzle as my warm body began to cool in the surprisingly cold water. After a few minutes, I wrapped up in a towel, grabbed a Baltika, and headed to the anteroom.

The rest of the crew lasted a surprisingly long time in the banya. One by one, they exited and headed to the same place I did: the pool. They, too, returned with more beer, and we had yet more discussions. I was beginning to get comfortable with this whole scenario by now and silently chided myself for my earlier feelings of distrust. After a few more beers, I was informed that I was the first to be treated by the 'Banya Master.' I had no idea what this meant, but by now, the Baltika beer had taken its effect, and my defenses were lowered. I was instructed to go into the banya without clothing or a towel and lie face down on one of the benches. Despite my inebriation, this set off my alarm bells and excited my flight gene. Again, I relied on my trust in Ioury and decided what was about to happen was what was to happen. So, I complied. It was easier to lie down in the extreme heat as the hot air was higher in the room. I noticed a pail of boiling water on the wood-burning stove, and some birch leaves hanging from the ceiling.

Oleg entered the banya wearing only a green felt hat formed in the shape of a stylish Fedora. This nearly toothless man with his odd hat on an overly large beer gut was a sight. I watched as he dipped the branches into the boiling water, held them there for a few seconds, and pulled them out. He began beating my backside with these branches. My slightly cynical mind started thinking of this as a uniquely Russian sadomasochism.

Each whipping of the branches stung, and the hot water on the skin prolonged the sting. Oleg went on like this for 2 minutes before rubbing the branches harshly into the skin. This was altogether a new form of pain. I questioned the sanity of these men at this moment. Next, Oleg commanded me to turn over, which was the real test of my faith. I decided to comply and was greeted with the same treatment as the backside. There was some solace in the fact that no particular attention was paid to any body parts, so I gradually became more comfortable with this situation. After about five more minutes, Oleg told me to go out and sit in the pool. As I exited, the team of guys looked closely at my face for expressions revealing my state of mind. I smiled, and that seemed to break the tension, but I was smiling at how insane this whole experience was.

After I spent 10 minutes in the pool, I grabbed my towel and covered myself as I lay under the shade of a large tree. I leaned against the trunk and drank some water to rehydrate myself. As I sat there, I noticed a familiar feeling: being stoned. Something was in those birch leaves, and it got into my circulatory system through the skin resulting in a new intoxication. It reminded me of my experience with hashish when I was 15 years old. It was not long before I fell asleep under the tree. I woke about two hours later, no longer feeling stoned but covered with bugs of various types and very uncomfortable. I took another dip in the pool but was hit immediately with what I called Lenin's Revenge: diarrhea. My Russian friends were still making fun of me and my apparent lack of robustness but were kind enough to make a tea from local weeds that they said would cure the problem. As I drank the tea, I saw the ants and insects floating to the surface. I drank it and ignored the extra protein in the spirit of open-mindedness. We ate a good lunch of shish kabobs, called 'shashlik' in Russian. After lunch, we returned to the dacha; by the time we arrived, it was 5 PM local time.

Alexander decided he was too intoxicated to drive back to Moscow, and the rest of us would ride the bus. I felt trashed entirely after this experience of being sunburned, still slightly high, filthy, and bug-bit. Yeah, this was fun. As we walked down the dirt road and waited for the bus, I realized many other Muscovites would return from their dachas for another week of toil for the state. As we boarded the bus, the stench of human armpit odor was striking. Most were standing and holding the handrails, making it even worse. Many were also drunk, mixing the smell of cheap alcohol and Cosmos cigarettes. I decided that this is what it must have felt like to be a 'real' Russian, and it brought many thoughts. We arrived at the subway station, and all parted ways. I couldn't wait to return to my plush Marriott hotel and sleep in its soft bed. However, this was an experience of a lifetime that I decided to keep entirely to myself not to give people the wrong idea about me. I kept this secret for many, many years until I told this to an AIAA dinner party, and it brought the table to unstoppable laughter.

We took our first trip to the launch site later that year. The area, known as Baikonur, is situated in modern-day Kazakhstan. It was long of interest to Western intelligence agencies, and few Westerners had ever been there. Francis Gary Powers had photographed this site on his last famous flight where after imaging Baikonur, he headed north to nuclear weapons sites at Chelyabinsk, where he was shot down. As Americans, visiting the site had meaning to us, and the gravity of what we would see was not lost on us. The purpose of the visit was to inspect the launch facilities in preparation for our satellite launch a few years away. The unmentioned purpose on our side was to see this once-forbidden city and military site and to test the willingness of the Russians to open up to us on activities and capabilities there.

The American delegation for the trip included me as the sole technical expert, our Program Manager Bruce Peterson, Len Caveny from the Missile Defense Agency, Col. Larry Ortega of the U.S. Air

Force, and a myriad of persons from the 'U.S. State Department' whom Len referred to as 'the people upstairs.' Whenever Len referred to people this way, it was typically a euphemism for 'spooks and spies.' When I asked several of them their roles, they consistently said they were there to 'assure that U.S. foreign policy was followed'. I smiled at this response and nodded my head.

Standing on an unfinished Soviet shuttle at Baikanour. Image Credit Jim Cantrell

Our trip to Baikonur occurred on a private jet. This Tupolev 134 was owned by the Energia Corporation, one of the largest Russian Aerospace firms and roughly the equivalent of Boeing. The delegation was about twelve people, and there was plenty of room for all of us. Getting to Baikonur on commercial flights was difficult and often involved refueling stops since it was usually a 5-hour flight. Having the convenience of the private jet made this a delightful journey. We took off from a private airport, Vnukovo 3, generally reserved for military brass and politicians in those days. When we taxied out on the runway, Boris Yeltsin was in front of us and on his way to some important meeting in Europe.

The plane was reasonably well-kept, and we even had cabin service. By the early 1990s, the Russians had started absorbing some of the lessons of Western capitalism and were emulating many of our business ways. One of these lessons was showing movies on the flight. In this case, they had installed a conventional TV from home in the cabin with a VHS tape player. The screen measured some 12 inches across its diagonal, and the sound left direct from the TV. After the flight took off, they began playing a Rambo movie starring Sylvester Stallone. The irony was not lost on us that an American delegation of aerospace experts and spooks was flying on a private plane to a once forbidden city to watch a Hollywood re-enactment of a proxy war between the Soviet Union and the United States being played out in Southeast Asia.

We landed in the city of Leninsk, which is officially part of the Russian military installation of Baikonur. The airport is a few miles away from the city, forming part of the Cosmodrome. The airport terminal was tiny and was not made for extensive traffic. As we landed in the late afternoon, the only other aircraft on the runway was a HIND military helicopter unloading a herd of goats and their nomadic herder. It was a strange scene indeed. A bus greeted us on the tarmac and took us to Leninsk. Although we passed through a small section of Kazakhstan, we didn't see any border guards or must-show papers of any sort. We continued out of the city towards the launch site where the first human was launched into space: Yuri Gagarin. This is one of two sites where the Soviet-era Soyuz rocket launches from at Baikonur. The Cosmonaut Hotel was located nearby and was where the astronauts normally stayed the day before launch. As it turned out, we were scheduled to see a Soyuz cargo launch to the Mir Space Station the following day, and we would be staying in the hotel next to the launch vehicle. The hotel is only a few miles from the launch pad, and you can smell the kerosene rocket fuel when it's being loaded into the launch vehicle.

Our first night in Baikonur was exciting, beginning with a tour of the Soyuz rocket on the launch pad. It had not yet been loaded with fuel, so we could walk up to the launch vehicle on the pad. It was an immense structure towering 150 feet tall and weighing over 750,000 pounds when fully loaded. To say that this was impressive would be an understatement. That we were some of the first foreigners to see these rockets up close was just as remarkable. However, one thing that stuck with me and would become a theme for most of my work in Russia was the crude workmanship of public works structures. The launchpad and its gantry, in this case, were very crudely welded and accompanied by jagged cuts along the edges of plate steel used for walkways. The welds were almost universally sub-standard and would never pass most basic Western welding standards. We were also stunned by the proximity of the launch control room to the launch vehicle. It was no less than 150 feet from the launcher but an underground bunker. One of the tour guides described what it was like to be in the bunker during a launch, and it didn't sound boring!

After the launch site tour, we returned to the hotel and had dinner with our Russian hosts. By now, I was used to the long tables with every manner of food on the table, along with bottles of vodka and shot glasses. We took our places at the table, and the immediate toasts began. We toasted the eternal friendship of the Russians and Americans, our project, our loved ones, our homes, and each other. Many of our U.S. government guests were utterly inebriated when the dinner was over.

On the other hand, I had developed a technique where I would substitute water for vodka in my shot glass without anyone noticing. In this way, I could maintain some level of sobriety and vouch for my U.S. colleagues who were becoming uncontrollably drunk. I half suspect that there was an underlying reason why the Russians did this, but mostly I think that it was simply to drink too much themselves. During this first dinner, I discovered that one of the Russian Colonels, Sergey or "Serge" as he liked to be called, spoke French. His English was sub-par, as was my Russian, so we began to talk in French. Serge developed a strong fondness for me and dominated my attention for the evening. He was a nice fellow, but he did manage to monopolize my time to the annoyance of my other Russian hosts.

The morning came early after a night of over-eating and too much vodka. Some of us had breakfast in the morning before we left to see the launch. After breakfast, the bus picked us up in the morning, and we headed out to the VIP observation site. Serge was still inebriated at this young hour of the day but filled with a bit less energy. The observation site was approximately 1 mile from the launcher, giving us a very clear view of the rocket. We smelled the kerosene in the morning at the hotel as they were fueling the rocket, and we could still smell it at the observation site, which was a bit closer. We all stood outside on the crisp fall morning and talked while the countdown continued. The crowd got quiet when it got to T-60 seconds, and many got their cameras out. We noticed a Japanese film crew, who had joined us the night before for dinner, was positioned halfway between us and the rocket. As the countdown ran under ten seconds, the crowd became silent as we watched the first stage engines light up. Since the flames ran down into a giant flame trench into the desert, you could only see the reflection of the flames on the side of the rocket. Still, the extreme brightness of those flames reaching my eyes tricked my ears into believing that I heard loud noises along with the bright rocket engines. The vehicle slowly lifted off the pad and the flames became more visible. I could now feel the infrared heat from the flames on my face as if I were going to get a sunburn. As the rocket rose, the wave of sound, which had been traveling much slower than the light, hit us with a thunderous roar. The rocket continued to ascend, and the noise became louder and louder. It became so loud that you could feel the vibrations in your internal organs. At about this point, I realized that I was in mortal danger should this vehicle fail, and there was little I could do to stop it if a failure happened. I continued hoping for the best. After about 40 seconds after ignition, the launch vehicle had ascended far enough that the sound began to subside. At this point, someone broke the silence in the crowd with an exclamation of "Oh wow!!!!!". Clapping broke out, and people began talking about what they had seen. It was a visceral experience that few of us expected, and I have never since been that close to a rocket launch.

Our next stop on this day was to go and tour the actual launch site where we would launch from. This was the second Soyuz pad about 20 miles from where this one launched. The roads on the Cosmodrome were rough, and the government-furnished bus would bounce with exceptional precision every 3 seconds. This soon caused many of us to have very full bladders that needed attention. Fortunately, when you are a male, all the world is your bathroom. Four stops on the way to the launch pad helped us rid our bodies of the morning coffee. The launchpad was very much like the one that we had toured the night before, except, not having a rocket on the pad, it was easier to inspect the totality of the installation. This time I could crawl all over the pad and examine any detail I liked. I started at the top and worked around the pad and into the large flame duct that emerged underneath the launch vehicle. The flame duct redirects several million pounds of thrust from the rocket as it ignites on the launch pad and sends it out horizontally onto the steppe. Water is mixed with the flames to cool the fire and absorb the immense acoustic energy. Standing in the middle of the flame duct, I could see the many large panels of poorly poured concrete stitched together to form something akin to a giant spoon. The concrete extended out into the desert for what seemed like a mile. We toured the rest of the launch complex seeing where the launch vehicles come in by rail, are assembled horizontally, and taken out the other door to the launch pad. It was an immense operation, and we would place our little satellite Skipper inside the final stage in between fuel tanks. On top of that, an Indian remote-sensing satellite would ride into orbit.

Author standing in the flame trench of the Soyuz pad used to launch Skipper from Baikanour. Image Credit Jim Cantrell

The next stop on our tour of Baikonur was the most interesting. We approached a large building that could easily house several small U.S. skyscrapers lying on their side with a building height of 200-300 feet. I asked Colonel Serge, who had appointed himself my best friend in the world, what was in the building, and he smiled at me. As we passed through the door and into the main interior of this immense steel building, we were greeted with the sight of four Soviet-era shuttles. The original, 'Buran' (pronounced Booran), greeted us. Buran had famously flown one mission and made a single orbit around the Earth before it returned to the Earth on 15 November 1988. We could see the heat streaks on the U.S. Shuttle-like tiles protecting its structure from the intense heat of launch and re-entry. Two other shuttles were complete but had never flown. A fourth was under construction and covered with scaffolding. Only a sole guard occupied the building; clearly, any construction on unit number four had stopped long ago. We were all stunned by what we were looking at. In the west, only the Buran was known to have had a successful flight and return. The other three models were unknown to most Westerners, even those specializing in intelligence matters. We stood in shocked silence, realizing we were the first Westerners to see this fleet of shuttles rivaling the U.S. shuttle fleet. What was eerie was how remarkably similar in shape and construction the Soviet and American shuttles were. I had long heard that the Soviets had stolen drawings of the U.S. shuttle fleet and based their design on that U.S. design. The Soviet shuttles that I saw before my eyes confirmed that rumor.

Len Caveny in front of the Soviet Shuttle Buran standing with the Baikanour Base Commander. Image Credit Jim Cantrell

Our 'official visits' to Russia that included U.S. military officers always included others whose roles were unclear. Their names were always easy to pronounce and were common Anglo-Saxon names. I learned not to quiz them very much, but when I did ask them whom they worked for, the response was almost always 'The U.S. State Department.' I may have been a little slow on the uptake, but after realizing that these people had no technical expertise judged by the questions they asked me, I began to understand that they were 'agency people.' Perhaps Defense Intelligence Agency (DIA). Maybe Central Intelligence Agency (CIA). One of these people was on the trip with us to Baikonur and went by the name Larry. Larry and I stayed close on the journey as he had many questions for me. He stood next to me, looking at the collection of shuttles, leaned over, and said, 'There are some people in the agency who would give their left nut to see the inside of that shuttle under construction.' I responded, 'Why don't we have a look?'. Larry demurred and seemed scared by the idea. I turned and asked my new friend Serge in French if we could mount the scaffolding and have a look. Serge smiled at me and waved me behind him as he moved towards the shuttle under construction. I waved Larry along behind me, and he hesitantly followed.

I followed Serge as he led the way up the scaffolding. As the three of us climbed the somewhat sketchy ladder up the side, an angry woman came out of the security booth and ordered us off the scaffolding. Serge yelled back at her, 'I am with the base commander, and these are our guests. They can see whatever they like'. The woman waved her hand at us as if to express disgust and returned to her cozy and warm guard shack. We quickly reached the wing area of the shuttle, and I noticed that the rest of the team was following. Larry was busy taking pictures on a 35mm camera and mumbled, 'These rolls are going directly to the agency'. It all seemed like fun and games at the time, and I continued to look around the shuttle by walking around the payload bay and sitting in the cockpit seats. We all looked around this fantastic piece of history and soon headed back down the scaffolding. As I left the building, Serge handed me a white tile from the unfinished shuttle. He told me in French, 'This one will never fly. You might as well have a piece of it.'

Soviet Shuttle #3 Never Flown at the Baikanour
Cosmodrome. Image Credit Jim Cantrell

Following the stunning visit to the shuttle storage area, we headed over to its launch pad. This was also the launchpad for the massive Energia launcher meant for manned missions to the Mir Space Station and Mars. The Energia shuttle was placed on the back of Buran for its launch, much like the U.S. Shuttle. All of this required a massive launchpad. Unlike the Soyuz pads I had seen, the flame trenches on this were enormous.

At the core, three canals exited radially, each 300 feet across. They were another 200 feet deep and had significant water at the bottom. This was all designed to deal with over 10 million pounds of thrust for the original lunar program. Sergey revealed that this original N-1 launch site had survived and was converted into an Energia launch pad. The steel from the N-1 site was removed, and the concrete was left intact. Two large rail lines split apart by 250 feet or so remained, and this was used to transport the vehicle to the pad horizontally. We were again standing on top of history. I remembered seeing declassified images of the pads taken by U.S. satellites during the Cold War and could make out several of the features I could remember. The weight of history hit me at this site, and it's when I realized that we're being given access to the inner sanctum of the Russian space program. Few Westerners had ever seen these places or seen the space hardware being shown to us. This was indeed Glasnost, or openness, like Gorbachev, the former Premier of the Soviet Union, used to say.

As we boarded the bus to head back to the hotel, the senior commander of the Cosmodrome asked the group of people their opinions for the evening. We had two choices. We could return to the Cosmonaut Hotel, rest, and then have another large dinner there. The second option would be to take the bus to Leninsk and shop, followed by a special Russian treat - 'banya.' There was that word again. My mind immediately returned four months earlier to the sunny summer afternoon outside of Moscow with Ioury and Oleg, the banya Master. Len Caveny looked at me and asked me if I knew what the banya was. I just nodded my head up and down in affirmation. Len enthusiastically turned to me, saying, "Tonight we do a banya!". I could only imagine how this would go.

The Skipper team at the Baikanour Banya.
Image Credit Jim Cantrell

Before the banya, we were given the choice to go back to the hotel or to do some 'shopping' in Leninsk, the small town where workers from the launch site lived. Col Ortega and I opted to shop, and my new friend Col Sergei insisted on going with us. He separated us from the shopping group and insisted we go to his home for tea. Sergei's wife was shocked when a full bird U.S. Colonel and I showed up at her door, and Sergei demanded that she make us all tea and serve cookies. It was late afternoon, and his wife was charming, while Sergei was very drunk and becoming belligerent.

Col Ortega and I were unsure what to do other than comply with Sergei as we watched his wife nervously serve tea and cookies. I felt horrible for her, but there was little that I could do to change the situation. After about 15 minutes of drinking tea, a heavy knock on the door arrived, and Sergei went to answer it. I could hear the general's voice, and he was not happy. He was telling Sergei that after he took us away from the group, our security people panicked and thought that we had been kidnapped, which could result in an international incident. The general came into the room where we were drinking and asked us to return with him after politely refusing tea. Sergei stayed at home.

After reuniting with our U.S. group, we returned to the Cosmonaut Hotel for some much-needed rest. Our colleagues had become very worried about us and were happy to see us again. We departed the hotel at about 8 PM to join the Russian group at the banya at another hotel in Leninsk. It was a much more civilized affair this time, with the banya in the hotel basement. It was a large room with a table, pool, and a sauna-like banya in the corner. We started by eating and drinking at the large table with several toasts. I remained perfectly sober during this banya experience as I knew that my U.S. colleagues, all of whom had high-level security clearances, would need me to vouch for their 'good' behavior when we got back to debriefings in the U.S. I will admit that it was amusing watching grown drunk men heading off naked into the banya and then afterward, stoned much as I had been the time before, splashing each other in the shallow pool. I stayed out of the banya this time around and maintained surveillance. Little was said the following day at breakfast, and we got ready to depart back to Moscow.

The rest of the Skipper development was straightforward. As we entered the hardware development and test phase, it became a normal spacecraft development. In Logan, the Skipper 'spacecraft bus,' as it was, started to take shape. The electronic components were located inside a pressurized container behind a re-entry dome made of thick aluminum. I had done most of the thermal calculations to determine how far into the atmosphere re-entry this dome would last, and it was a fantastic experience to see this come together into reality. The Russians, meanwhile, were assembling and testing the propulsion system. We carried some 45 kilograms (100 pounds) of hydrazine rocket fuel, essentially making the entire satellite a flying fuel tank. By June 1995, the propulsion system was delivered to Logan and mated with the satellite bus. We had a minor drama when we discovered that the Russians used a positive ground on the chassis, and we used a negative ground on the chassis. It caused the power supplies to trip immediately after we mounted them together. After a quick evaluation, the Russians made some changes to the grounding strategy of the power system, and we had complete compatibility. The rest of the development focused on system-level testing and flight operations preparation. We had an unusual flight operation to implement since we used the Air Force Satellite Control Network, which had the control center in the famous 'blue cube' out at the Onizuka Air Force base. This was the nerve center of all U.S. military satellites, including the most classified spy satellites.

The question of allowing the Russians admission into this building was addressed many times, and the answer was always no. Instead, we set up a Mission Operations Center in Logan where the Russians would be with most of the U.S. team. I would lead the operations within the Blue Cube, take command plans developed by the combined team, and upload them to the satellite. The Russians found this whole plan very suspect, and the inevitable accusations were eventually made that the U.S. would 'fake' the data given to the Russians. Old habits from the Cold War die hard, it seems.

We shipped the completed Skipper to Moscow in October of 1995. After facing some last-minute export control issues, which were far from the ordeal they are today, we managed to ship it directly to Moscow Aviation Institute. They would then arrange to deliver it to the launch site, where Lavotchkin would take over and get it to the integration hall at the launch site. By early December, this would be followed by spacecraft checkout, fueling, and installation into the fourth stage. Len Caveny chose me to lead the launch crew at Baikonur and began planning for this responsibility. As it turned out, the spacecraft's shipping did not go as planned. Unbeknownst to our team, things had started tragically wrong on the way to the launch pad. The spacecraft had made it to Moscow and through customs in Moscow quickly enough. As undeclared equipment, MAI had arranged for the spacecraft to fly to Baikonur on a private cargo flight. The Lavotchkin crew was waiting to take possession of the vehicle once it landed on the old shuttle landing strip at the Baikonur Cosmodrome. However, we needed to prepare for what would happen as Skipper crossed the Cosmodrome to the launch site.

I was responsible for doing final checkouts on the spacecraft and integrating it inside the fourth stage of the Molniya rocket. A commercial Indian remote sensing satellite IRS-1C was mounted on top of us, so our integration happened first. We had carefully prepared and boxed up our Ground Support Equipment and tools, which were sent alongside the satellite. Once the equipment and satellite were confirmed to have passed through customs in Moscow, we set off on a plane to join the Russian crew at Baikonur. The first wave of the crew was me and a fellow employee named Jay. We took the routine flight from Salt Lake City through New York to Moscow. The plan was to arrive in Moscow and spend a few days there before we headed to Baikonur. We would stay in the usual Marriott in downtown Moscow and work on the final planning steps before flying out to the Cosmodrome. The flight from Moscow to Leninsk was another 5 hours, so it paid to spread this into smaller sections.

Our flight to Moscow was uneventful, and we arrived late afternoon on an early November morning. The snow had come early to Moscow, and we landed on a snow-covered runway. As we taxied up to the terminal at Sheremetyevo, the familiar feeling of dread began to well up inside me. Dread of the cold dismal weather. Dread of the dark and dingy passport lines that ran through the basement of the airport. It was so dark in the passport area that the only light came from the passport control office, where the officers examined and stamped the passport. They seemed to hang softly in the darkness like a floating globe. I dreaded this part of the trip but tried not to think about it.

As we disembarked from the plane, the stewardesses said their goodbyes, and we exited the plane into the well-lit upper part of the airport arrivals. To my surprise, a young man dressed in a black suit and tie held up a sign with 'Cantrell and crew' written on it. Jay Ballard and I looked at each other and assumed this was some red-carpet treatment. As I identified myself and Jay to him, he asked for our passports which we gladly gave him. He led us out of the departure areas, and we skipped the regular passport lines buried in the basement. We headed through baggage claim, grabbed our bags, and entered the cold Moscow fall. Winter was biting at our backs much as it had done to the Germans in 1941, and we were happy to get into the warm van waiting for us. We were quiet as the van exited the airport and headed into Moscow. Jet lag is a funny thing. You feel pretty good when you first wake up on the airplane upon landing. But the problem is that you start to feel very fatigued after an hour or two, and your mind doesn't work as clearly as it usually would.

This was about the point in our trip where our minds were still apparent, and I noticed we were not heading toward our usual hotel in Moscow. Instead, it seemed we were heading onto the city's northeast side. I asked the driver, in my patois Russian, where we were heading. He was noncommittal in any response. I began to worry but didn't communicate my worry to my travel partner Jay. He was not the kind of person you communicated worries to because he was naturally worried about everything anyway.

About 20 minutes later, I discovered that I was right. We were no longer going to our hotel. We arrived at a very tall apartment tower. Gary Popov was parked in a car outside of the complex. As soon as we left the van, Gary got out of his black Volga, a high-end vehicle reserved for VIPs and KGB, and greeted us with warm, open arms. Gary was always smiling in a typical politician-type way. Instead of embracing him, I asked Gary, "What the hell is going on?". He explained that there was an "irregularity with our visit" and that we would stay with his good friend Nadehzda. I asked Gary what the nature of the irregularity was. He was vague in his response to me and explained that we would be very comfortable here with Nadezhda until this problem could be solved.

We headed into the apartment and were greeted by an 80-something-year-old woman who reminded me very much of my grandmother. I could smell that she had something delicious cooking and awaiting us. She and Gary spoke in Russian for about five minutes while we unpacked our items into our bedrooms. It was very unusual for a Moscow apartment to have this many bedrooms, so it was self-evident that Nadezhda had been a significant person under the Soviet system. We were invited to sit down for lunch, and Gary bid us goodbye as we began to eat. Nadezhda fussed over us and the lunch and repeated over and over in Russian 'Oching V'Koosnee' (very tasty in English).

My Russian was modest, but I could communicate basic things to Nadezhda. I thanked her for the meal and for having us in her apartment. She was a very polite, kind, and caring woman. That much we could tell. We were still trying to understand why we were here. However, the excellent food and the warm rooms made us quite happy, and we were not inclined to complain. We finished our lunch and slept afterward, catching up on what amounted to middle-of-the-night sleep. I woke later in the afternoon and began wondering again what was happening with our situation. Being a Sunday, there were not many people around that we could readily talk to, so I resigned myself to reading and getting used to my new environment.

In the evening, Nadezhda cooked us dinner and headed out the door afterward. As she was leaving, she explained that she would be back in the morning, and if we needed anything from the kitchen, we were free to get it during the night. As she left, she closed the door and locked it from the outside. I checked the door after she left; oddly, I could not open it inside. There appeared to be locks that would operate only from the outside. I begin to get suspicious at this point. Nonetheless, I managed to get some sleep that night and waited until the morning when our new mother returned.

The arrival of the morning was mercilessly slow. I routinely woke at 0300 when I arrived in Moscow and could not get back to sleep. At 5 a.m., Nadezda reappeared at the front door unlocking the same locks that imprisoned us all night. She quietly snuck into the kitchen and began cooking. I could hear her singing quietly in Russian, and it was a beautiful song. I enjoyed this moment listening to her sing and felt a genuine peace. I was still very anxious about our situation. I wanted Gary to contact us today so we could better understand what the problem was. After all, we had a satellite waiting for us in Kazakhstan and needed to get there to test it and integrate it into the launch vehicle.

Around 11 o'clock, there was a loud knock on the door. A man called himself Kulikov and identified himself as working for Gary. He was there to pick us up and take us to the Moscow Aviation Institute. We prepared to walk into the Moscow winter with our thick coats and hats and went downstairs with Kulikov. We got into his black Lada. A driver took us to Moscow Aviation Institute, and we arrived at a very familiar building where I had been before negotiating the Skipper contract. As we went inside, we entered Gary's office and waited for him to arrive. Gary finally arrived about 30 minutes later, along with several of his colleagues who sat on the other side of the table. Gary gave us more details on our situation.

Nonetheless, the authorities in charge of the cosmodrome had yet to get the memo and had determined that something nefarious was happening. According to Gary, we were the recipients of the fury that went up and down the chain of command in the MoD. The short version of the story is that we were officially accused of espionage and were thus placed under house arrest until the situation could be resolved. As evidence of our house arrest, our passports had been confiscated, and we were in residence with Nadezhda until further notice.

I found myself very unhappy with this explanation. However, it was good to know what we were up against finally. It was hard for me to be angry with Gary as I could see this was also very hard on him. Somewhere in getting the Moscow licensing, Moscow Aviation Institute had not done its job. Perhaps they hadn't paid a bribe, a 'fee.' I recall one situation where Gary asked us to help deliver what amounted to a bribe and we declined as this is illegal in the United States. Perhaps it was something quickly resolved with a little bit of money on the part of Moscow Aviation Institute, or maybe it was something more complicated. I remained optimistic that this could be resolved in a day or two, and then we could head off to Kazakhstan and get our satellite ready for flight. This was early November, and we still had at least a month before it was scheduled for flight.

The post-Christmas flight schedule was dictated by the Indian satellite that was the primary payload, and we were simply a secondary payload beneath it. We needed it to meet their schedule. After talking a bit longer with Gary, he mentioned that while we could not have our passports back, Kulikov would be our guide around Moscow should we like shopping or need to get out and do anything. We were to remain with Nadezhda, and she would take good care of us.

We returned to the apartment later that afternoon and were greeted by Nadezhda, who was happy to see us again. This time she took some time to sit down and talk to me and was very tolerant of my elementary Russian. She had been married to a Soviet general who was a hero of the Soviet Union resulting from World War II actions. He died some years earlier from a heart attack but not from the Germans. It was evident that Nadezhda still loved him very much. We talked about how they hunted together and how much she loved spending time with him. I talked about my own hunting experiences, and through this, she and I found a personal bond. I didn't know what to make of Nadezhda. However, I didn't believe she was something evil or doing anything other than what she was asked to do. Jay and I did feel like prisoners, but on the other hand, we felt grateful to be with Nadezhda. I tried very hard to think of it that way rather than as being held under house arrest as Gary had qualified it.

When the evening came, I got on the phone in the apartment and called back to the Space Dynamics Lab to talk to my boss Alan Steed. I explained to him the situation we were in and that we would need help to try and resolve it. I asked Alan to get ahold of the US Embassy, and they could intervene on our behalf and at least get our passports back. Alan said he would do what he could and give me more information if I called him the next day. Alan was a good man, but I wasn't sure if he was up to this task or took it as seriously as I hoped. It was a dire situation, as our liberty was at stake. I called back the next day, and Alan had been unable to get any response from the US Embassy. On Wednesday, I heard back from Len Caveny, who called and talked about the situation. He mentioned that he and

Gary Popov had been in communication and would be personally intervening with the US Embassy and the Russian MoD to see if we could resolve this situation. If that happened, Jay and I could head to Kazakhstan to finally test and integrate the Skipper satellite and return to the US afterward.

I didn't hear back from Len Caveny for about a week after that phone call. In this period, Jay and I both started to become emotionally unstable. From a perspective 25 years later, it's interesting to look back and see how we dealt with such a stressful situation. One day I would be strong, and Jay would have an emotional breakdown. The next day he would be strong, and I was the one that would have the emotional breakdown. It was as if we depended on each other to calm the other down when we each had our emotional breakdowns. I think we both realized the seriousness of our situation and gradually, it took a toll on our psyche. We finally heard back from Len after another few days, and he mentioned that he was getting on a plane and coming to see us. We had been in Moscow for about ten days, and this situation was not resolved.

I did call Gary and asked him the next day to see what he knew. Gary needed something to update us on. He did offer to take us to dinner or to have somebody escort us shopping if we were bored. I politely declined his offers. That evening I called Alan again. Alan seemed unable to do much for us and mentioned that he had been talking to the Missile Defense Agency and Len Caveny. This was my day to be emotionally unstable, and I became very angry with Alan on the phone. I called him names that I would never have called decent and honest people. However, I felt like I was in prison, and my boss was unable or unwilling to do much for us. I later apologized to Alan, and he did not hold my outburst against me. Rather, he seemed to think that it was moderate, given the circumstances.

When Len Caveny showed up on Friday, he came to see us at the apartment. It was terrific to see an American there and learn what was happening with our situation. Len, a long-time Navy man, was a fighter. I could see it in his eyes that he cared about our situation and that he would do his very best to resolve it. This lifted our spirits, and we look forward to a speedy resolution. As Jay and I spent the weekend shopping for little Russian trinkets and toys for our children back home, we knew we would be coming home soon. It was a nice weekend, and my friend Ioury invited us to his apartment for dinner. During the dinner, he told me he was learning about our situation from some Russian Ministry of Defense friends. I could see it was a more severe, dire situation than we had imagined. According to Ioury, we were being 'accused of espionage' even though there was no basis for such an accusation. Ioury also mentioned that he had friends in the new Russian media and would go to them if this did not resolve itself quickly. I asked him not to do something crazy as I worried about his well-being. I told him that Len Caveny had come to Moscow to resolve the issue and that I was optimistic he would do so.

Over the following week, Len spent time with the US Embassy and the Russian Ministry of Defense. He reached very high into the Ministry of Defense (MoD) and met with General Valery Mironov, the former commander of Soviet forces in the Baltics. However, Len could not get any agreement to let us continue our work in Kazakhstan on Skipper or go home. Talking to the MoD made it unclear to Len what the problem really was. As if to add to our misery and a symbolic slap in the face, my own United States embassy was unwilling to help. Although we briefed the Moscow defense attaché at the Embassy on nearly every visit to Moscow, we made, the State Department treated us as if we were non-official cover spies (or NOCs, as the parlance went). What that meant was that we were not to be acknowledged and not to be helped. This made me very angry as I felt abandoned by my country in my time of need. Len went home after a week and a half in Moscow empty-handed. I was beginning to become emotionally fragile again after nearly three weeks of captivity.

The Monday following Len's departure, I began calling Alan Steed and complaining about the lack of action. However, this time when I called Alan, he mentioned that one of our board members Jake Garn, a former Utah senator, and former shuttle astronaut, was intervening on our part. He had known Al Gore, then the US vice president, during his time in the Senate. Jake and I knew each other from SDL board meetings because of our mutual love of cars, and we often discussed cars and airplanes during the meeting breaks. Alan relayed on Monday that Senator Garn had called Al Gore to seek his intervention on our behalf but has not yet heard back from him. This gave me again more optimism. However, we have not resolved our issue and have been in Moscow for almost a month. We were getting very close to when we had to leave to take care of the satellite waiting for us in Kazakhstan or launch without the satellite aboard.

Good news came a week later when Alan called reporting that Al Gore had returned Senator Garn's call and that we would be put on as an agenda item between Al Gore and Victor Chernomyrdin, the Prime Minister of Russia, during the next bilateral call. That was to occur the following week. If all went well, we would get our passports back and be free to travel to Kazakhstan. By this time, we had exactly two weeks left until Skipper needed to be integrated into the Molniya rocket or miss the ride to space.

The good news came a few days later. Gary Popov informed me that the situation had been resolved and that we would soon receive a letter from Prime Minister Chernomyrdin along with our passports. It was early on a Tuesday morning after breakfast that there was a knock on our apartment door. Nadezda answered it, and I heard some soft Russian spoken and recognized Kulikov's voice. I went to the door, and as I did so, Kulikov extended his arms to me, holding a letter as if it were white hot. I grabbed the letter from him and saw a wax seal on the envelope that was the official seal of the state of Russia. Kulikov also handed us our passports back. He smiled and said he would accompany us to Kazakhstan when we departed. He also mentioned that Moscow Aviation Institute was making the arrangements for us to go to Kazakhstan and that we should be ready to depart any day.

I opened the letter after breaking the wax seal and read it. The Russian was very formal it was challenging to read. However, I saw that we were explicitly named as guests of the state and that this letter instructed anyone who encountered it to allow us the full privileges as guests of the state of Russia. It appeared that the Senator's intervention on our part had bought our freedom. After we bid Kulikov goodbye, I took my passport into my bedroom and quietly wept. I knew we had come dangerously close to something I never wanted to encounter again. Fate is a hunter; for now, I could not dwell on my close brush as I had a mission to complete.

Getting to Kazakhstan was much more complex than we had ever imagined. Russia is a huge country. Flying from Moscow to Kazakhstan takes about six hours, and there aren't many commercial flights between Moscow and Leninsk, the civilian city next to the Baikonur Cosmodrome where the Russian rockets are based. The first time we went down there with the US delegation, we took a private plane owned by Energia. That was no longer available since we didn't have US flag officers with us this time. Commercial flights to Kazakhstan were only available on Tuesdays and Thursdays, and we missed the Thursday flight. We cooled our heels in Moscow for the time being while we tried to find other ways to get there sooner. The only real option was to leave and arrive late Tuesday afternoon. The spacecraft had to be integrated by the end of Wednesday, or we would forfeit our flight with the Indian IRS-1C. Tuesday came, and we flew to Kazakhstan with our FSB minder Kulikov. The flight was uneventful, but we landed late at night, around 11 PM. A Russian government jeep greeted us at the Leninsk airport staffed with government representatives. As it turns out, there was a small strip of Kazak territory between the airport and the city of Leninsk that we had to pass through.

As we passed through this narrow strip of another country, there were border guards, and they were not in a good mood. These young Kazak soldiers were dressed in full battle fatigues and were carrying loaded AK-47s. We stopped at the border, and the Russian representative spoke Kazak with the guards. They asked for our passports in a very aggressive manner, and we gave them over reluctantly. An argument erupted between the border guards and our Russian government escorts as, apparently, we needed a Visa to cross this part of Kazakhstan, and the border guards became very agitated when they discovered we didn't have the visa. As the argument intensified between the soldiers and our Russian hosts, one of the border guards pointed his AK-47 directly at my head. At the same time, he spoke to the Russian government representative in a very menacing way. Not understanding a word of Kazak, I had no idea what was said. However, after a few long minutes of having a loaded AK47 pointed at me, they resolved the matter, and the gun was removed from my forehead. We were given our passports back, and I felt like we had again cheated fate's hunt for us. When the gun was pointed at me, events seemed to move very slowly, and my mind raced quickly. I began to think that this was where I would die, in the middle of nowhere where I didn't know anyone. I imagined that friends and family would wonder where I had disappeared to. To say that this was an overwhelming situation is, to put it mildly. I was thrilled that we were finally on our way and tried again not to be distracted by the thoughts of what had just happened.

We arrived at the integration hall at midnight, where Skipper awaited us. The Moscow Aviation Institute provided meals and other items to help us work throughout the night. Our colleagues from Lavotchkin had unpacked all our crates and laid out and set up all our ground support equipment to test out the satellite. Jay and I had had plenty of time in Moscow while we waited to go over all the procedures that needed to be done to test the satellite and had them practically memorized. The testing went very quickly and smoothly, and everything worked well. We tested the solar panels to ensure they generated electricity, and we tried all the internal functions of the satellite, which operated flawlessly. The trip from Logan, Utah, to Kazakhstan had done no damage to the satellite. We declared the satellite ready to launch at about 5 AM. The following day it would be fueled and installed onto the fourth stage. I had managed to smuggle in a satellite phone system from the US so that I could call back to the US and get authorization to launch. This was a formality of our US system, and I called the Program Manager, Bruce. I told him we had completed the integration testing, and the satellite was ready for launch integration. I received approval to proceed.

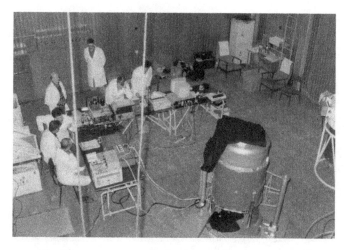

The Skipper team at the Baikanour Integration Hall.
Image Credit Jim Cantrell

At the end of all this drama, I was presented with a can of vodka from my colleagues at Lavotchkin to commemorate all we had been through. This can of vodka, looking much like a Coke can, had an image of a skull on it and the words "death vodka" on it. I drank it very quickly, much like you would a soda, and quickly became very sleepy. I was given a sofa to sleep on but was awoken several hours later by the Indian crew who arrived to work on their satellite. It was almost surreal waking up and seeing these Indian engineers looking at me very close to my face. I had no idea where I was for a minute and even wondered who I was.

The Skipper team with the ready to fly Skipper satellite.
Image Credit Jim Cantrell

We had a plane scheduled to return to Moscow that afternoon, so I found my Lavotchkin colleagues and witnessed the satellite being fueled. After that, we returned to the Leninsk airport and waited for the plane. The plane was to arrive at 2 p.m., but it was late arriving. We waited in a tiny room with no chairs or seats, with approximately one hundred fellow Russians. After waiting a few hours, the Russians became unruly and began drinking. After another hour, small fights started to break out. By the end of about four hours, I was one of the few sober individuals in that room. Finally, at about 7 o'clock, the plane arrived, and it was not the normal Aeroflot plane. Instead, a charter aircraft named "Vanukova Airlines." We didn't care much about why the previous plane had caught fire by this time. This new replacement plane and just boarded the flight. As we were boarding, I could hear the pilot berating passengers on the loudspeaker and telling them, "If you were too drunk to climb the stairs to get into the aircraft, you're not going to be allowed on this aircraft to return to Moscow." We found some seats further back in the plane and sat down. A lady across from us was bringing chickens home with her and sat in the aisle across from us. A man sat in front of us with a massive suitcase on his lap. It was all very surreal but par for the experience thus far.

As the plane took off, the Catholic deep inside me thanked God that we were finally leaving. I quickly became preoccupied with the reality that the seat in front of us was not bolted down and was falling into us under the acceleration of takeoff. Jay and I both reached up to press on the seat to hold it forward to prevent this heavy man with a suitcase on his lap from crushing us. Ceiling panels from the airplane began to fall in the aisle as the plane took off, and the stewardess nonchalantly put them back in place as if this was normal. I fell asleep quickly and was woke by the plane landing in Moscow many hours later. As we got off the plane, I kissed the ground. I was so grateful to have survived everything we had been through. However, it was time for us to head back to the US and get ready to fly the satellite.

Our crew from Lavotchkin remained in Kazakhstan during the launch. Flight operations in the US were divided into two parts. Flight planning and data analysis were in Logan, Utah, and the actual commanding of the satellite was done through the Air Force satellite control network with its headquarters at the blue cube and Onizuka Air Force Base in Sunnyvale, California.

The "Blue Cube" at Onizuka Air Force Base where Skipper mission operations occurred. Image Credit US Air Force

I had missed Christmas while I was gone, and I could scarcely tell my family what had happened. The Missile Defense Agency had briefed me that this entire episode of being held under house arrest would be treated as "classified information," meaning I could not discuss it openly. My family was suspicious that I had a girlfriend in Russia, but that could not be farther from the truth. Unless, of course, you might consider Nadezhda my girlfriend in some ways!

On the evening of 27 December, I briefed the base commander on the mission, and we assessed readiness to support it from a ground station perspective. We got clearance from the base commander to move ahead, and the mission launched at 11 PM California local time. The launch went off without a hitch. It was picture-perfect. Our first contact was over the Indian Ocean, and the spacecraft responded perfectly as we commanded it to spin up and assume its standard orientation. All was well with Skipper. We continued the mission throughout the night and configured Skipper to be ready for its forty-day mission.

About 20 hours into the mission, however, we began to see the battery voltage dropping. This was not a good sign and meant that we were not receiving power from the solar panels as expected or that power consumption was more significant than planned. We didn't understand why this was happening at all. Indications from the spacecraft indicated that the solar panels were creating power, and the net power consumption of the spacecraft was well characterized and within limits. As a precaution and in line with our contingency planning, we began to shed power loads to minimize power draw and hopefully remedy the situation temporarily so that we could buy time to understand it. Despite some aggressive load shedding, the battery voltage continued to drop without any reasonable explanation. We managed to operate the spacecraft for another thirty-six hours until disaster struck.

Skipper launch from the Baikanour Cosmosdrome.
Image Credit Ioury Bojor

The spacecraft came up over the Hawaii station about thirty-eight hours into the mission. Its vital signs were not good. The battery voltage was very low and was reaching a critical state. As the flight operations manager, I had to make snap decisions on what to do from the Blue Cube at Onizuka Air Force Base. The loads we had shed comprised nearly everything in the spacecraft not vital to remaining in a communicative state, and we had very few additional loads we could shed. More critically, the propulsion was starting to get cold enough to freeze the rocket fuel aboard (hydrazine), which freezes at about the same temperature as water. Based solely on the danger of freezing the propulsion system, I turned on the thruster catalyst bed heaters to warm the feed tubing where the freezing would first occur. As we executed the command, the spacecraft went utterly dark, and we lost the radio signal. This meant we had brought the batteries down to a point where nothing worked anymore, and the computer and telemetry system had shut down. I knew at that moment that we had lost Skipper.

As a result of this event, we declared a "spacecraft emergency," which gave us top priority on the satellite communications network. On a network like the Air Force Satellite Control Network (AFSCN), where many operational defense satellites are using the downlink facilities at various levels of priority, this is a bit like pulling the pin on the hand grenade, tossing it into the network planners' room, and shutting the door. In any case, we weren't sure if the spacecraft was completely dead, and we needed a complete set of resources from the US Air Force to determine where it was and what was happening with it. Suspicions on both sides had characterized the entire Skipper program, and by now, the US Air Force officers were openly discussing the possibility of sabotage. As a result, many classified resources are called into play to image the satellite and to understand if it is still operating despite its inability to communicate with the network.

We received the requested priority on the network and tried to command the spacecraft back every time it passed over a ground station. We were ultimately unsuccessful in getting any response from the spacecraft. After 48 hours into the mission, ground radar stations started picking up large pieces of debris surrounding Skipper. This was corroborated by the Russians, who also had a "headcount" of debris from their space radar tracking network in the same place where Skipper once was. It appeared that not only had Skipper ceased to operate, but something that happened to cause it to break into pieces. We were not at all sure what happened based on the data that we were seeing. I was very depressed by this point and frankly felt after all I'd been through. This was just the bad luck of what was a lousy luck mission.

I returned to Logan a day later to participate in a failure analysis. It was evident by this time that Skipper was no more, and what was left of it was in pieces in orbit. One of our crack engineers, Brent White, discovered the "smoking gun" evidence one afternoon through the telemetry data. As a spacecraft passed from the dark side of the Earth and into the sunlight, referred to as the "Umbra Exit," the battery voltage dropped. Usually, the voltage would rise when the solar panels are illuminated, as this rise in voltage charges the batteries. The only conclusion we could have based on these data was that the solar panels had been hooked up backward. Somehow the solar panels were discharging the battery rather than charging the battery. Usually, a diode system would prevent this discharge, but this was not the case with Skipper.

With this evidence in hand from the spacecraft telemetry, we studied the diagrams of the power system and concluded that either the system was designed with the power system wired backward from the start or the safety plug, which disconnects the solar panels from the batteries, installed right before launch, was somehow wired backward. My testing records at Baikonur indicated that the solar panels provided positive power with our ground support equipment, and currents flowed in the correct directions. And while the safety plug was out of the spacecraft during this test, we put it on the ground support equipment to check that we were getting positive voltage increases from the solar panels with artificial lighting. Our only conclusion was thus that somehow somebody had reversed the safety plug wiring after we had left using a different safety plug..

$7 million USU satellite wired wrong

LOGAN (AP) — Utah State University scientists have confirmed that a $7 million joint U.S.-Russian satellite failed because Russian scientists wired a battery charger backwards.

The satellite, known as Skipper, was launched Dec. 28 from Kazakhstan on a 30-day mission intended to help detect and identify incoming missiles.

USU and Russia worked on the mission for more than two years after a contract was awarded through the Defense Department's Ballistic Missile Defense Organization.

But the 550-pound, 59-inch satellite went suddenly quiet less than a day after's its launch and since then scientists in both countries have been trying to find out why.

James Cantrell, main engineer for the mission at the USU Space Dynamics Laboratory, which operated as mission control for Skipper, said solar panels that were to recharge the satellite's battery were connected backward.

"They were hooked up in reverse polarity and had the effect of discharging the battery instead of charging it." Cantrell said.

"It would be like wiring your alternator backward," he said.

Cantrell said the power unit was not designed to catch the error.

"It's always the simple stuff that kills you," he said. "It should have been caught in the design stage but wasn't."

Sen. Orrin Hatch, R-Utah, confirmed that the satellite's power unit was built in Russia by Russian scientists.

"Unfortunately, the $7 million U.S. investment in Skipper was

"They were hooked up in reverse polarity and had the effect of discharging the battery instead of charging it. It would be like wiring your alternator backward."
— James Cantrell, main engineer for the mission

lost when the satellite's power plant failed," Hatch said.

The solar panels aboard Skipper were to keep its nickel-cadmium battery charged during the mission.

But, because it was wired backward, the recharging system began sucking juice from the battery the minute it was launched. Cantrell said.

Cantrell said the solar panel board and propulsion system were designed and built by a Russian government-owned organization equivalent to Boeing with a long history in aviation.

Another agency made the battery, he said.

"One company made the solar panels and one made the battery and they didn't carefully look at the solar interfacing and when the interfaces were backward, nobody caught them," Cantrell said.

"It's not that they're stupid," he added.

"With all the testing systems everything looked good. We checked it in Logan and the measuring system aboard the satellite was not such that we could detect the error."

Unfortunate press article on Skipper quoting the author's off the record comments. Image Source Newspapers.com

Given the hysteria that I had experienced during my house arrest, and the fact that the Russian MoD was convinced that we were spying on them, my colleagues in the Missile Defense Agency began to wonder openly about sabotage. This would have been a perfect way to do it. None of our Russian team admitted to having any role in this. However, had they been brought into such a scheme, they would never have been allowed to discuss it. The Missile Defense Agency came up with an official conclusion that the Skipper had been sabotaged and that it had been ultimately motivated by a lack of trust between the two countries. One of the founding objectives of Skipper was to develop the confidence to base more prominent and more extensive joint defense programs. The thinking at the time was that this would, long term, keep former Soviet scientists from leaving to all the problematic third-world countries with nuclear weapons and ICBM programs and would, so we hoped, draw the United States and Russia closer. The Skipper catastrophe set that whole program back by at least a decade. I felt very burned by all of this and wanted little to do with the Russians going forward, given my experiences.

After this job was over, my boss Frank Redd gave me a new job to lead business development for the Space Dynamics Laboratory. In essence, I oversaw sales for the organization. This was a new area for me, and I was very enthusiastic to leave Skipper behind me as I put together a great team of people and began pursuing new programs for the lab. While I began to forget about Skipper, the rest of the world had not. Six months after Skipper's demise, a local journalist had heard a rumor that we had hooked up the solar panels backward and called me to verify the story. The journalist was known to my family, and I agreed to talk off the record to set the story straight. This is where I had my big lesson in not trusting journalists and the promise to keep conversations "off the record."

As I spoke to the journalist about this problem, I framed the situation in the light that 'we weren't idiots and had tested the system before flight". I continued to explain that somehow the solar panels had become hooked up backward after we finished testing the satellite, and probably this was through the Russian-supplied safety plug. I had to keep many conclusions to myself but tried to frame them favorably to the team without that information. While my new friend had agreed to keep this conversation, and more importantly, my identity, off the record, he did not. It appeared in our local Harold journal and was picked up by the Associated Press. The AP headline was "$7 million USU satellite wired wrong".

Unfortunately, at the same time, my comments appeared on the national news, US astronaut Shannon Lucid was getting ready to fly to the Russian space station Mir. The Russian astronauts colorfully commented to the press that they were "happy a woman was coming to the space station because it needed some tidying up." This blatant sexism set the world on fire, and my name got drug into the overall story to illustrate how stupid the Russians were by hooking up the solar panels backward while being backward knuckle-dragging bigots themselves. As this showed up on the front page of the Washington Post, our masters at the Pentagon were very unhappy with me. They called my boss Frank Redd with instructions to fire me because I had talked to the press and violated many of the rules of engagement. Frank was furious with me. However, after I explained what had happened, he understood better and decided not to fire me. He was still not happy with me, and I knew it. I did manage, however, to keep my new job, but my pride was very hurt. While this was my first lesson and dealing with the press, I've had excellent relationships with them, and off-the-record topics have stayed off the record since then.

For the next few years, life assumed a sense of normalcy. I worked sales at SDL, watched my children grow, and began to build a house I had long dreamed about. In the same period, sometime in 1997, my old friend Lou Friedman called me out of the blue. I had not heard anything from Lou for many years, and it was good to hear from him again. He had an unusual excitement in his voice and mentioned that he had interest from a group called Idea Lab to fund a private solar sail mission. His idea was to have our friends in Russia at Lavotchkin build the solar sail and launch it and Idea Lab would fund it as a commercial venture. Idea Lab was an Internet company incubator, and while I could not see the commercial link, I was most happy to help.

Lou Friedman is considered the modern father of the Solar Sail. He had advocated for a solar sail mission to Haley's Comet while at JPL in the 1970s. That mission never happened, but Lou never gave up on the idea of a solar sail. He eventually left JPL, finding it a stifling slow environment, and founded the Planetary Society.

As Director of the Planetary Society, he always sought somebody to fund his solar sail dream. The Gross brothers, who founded Idea Lab, were his latest hope that this might happen.

Lou called me to see if I would go to Moscow with him and a few others to discuss this idea with the Russians and the Gross brothers. I, of course, was willing to do almost anything for Lou and agreed to go. Our trip to Moscow was in the late summer and involved Tony Spear, the Mars Pathfinder program manager, Lou Friedman, Rex Ridenour, and me. The Gross brothers also came along with us on this trip, and they had some other meetings besides ours but accompanied us to meet with Lavotchkin. The trip was ultimately unsuccessful in getting interest to fund the Solar Sail. Still, it ended up creating of interest with Lou Friedman to find other private funding for the Solar Sail.

After this meeting in Moscow, we will pursue other funding paths for a private mission. We began talking with Carl Sagan's widow, Ann Druyan, who ran a Cosmos Studios group. She began to be intrigued with this idea of a privately funded Solar Sail mission, especially one that would be tied to using former Soviet military capabilities such as converted ICBMs. She thought of it as a "swords to plowshares" kind of demonstration, which would be something that Carl Sagan would have been proud of. These discussions started slowly, but Lou was able to find money to do a television documentary on building such a solar sail mission with the former Soviets and launching it on a converted submarine-based ICBM. We eventually engaged the Arts and Entertainment network, which was willing to underwrite this project. The solar sail mission would be called Cosmos One, flying out of a Soviet submarine on a converted Soviet submarine-launched ICBM. Our friends at Lavotchkin would build the solar sail spacecraft, and the Planetary Society would serve as the managing agent for the project. Lou asked me to be the project manager, and I agreed. We began this program in early 2001, and most of the work was with the Russians initially. I was deeply involved in managing it and making a few trips to Russia that year. My work on Cosmos One was about to lead to an event that would change my life forever when I got a phone call from somebody about changing the world as we know it.

CHAPTER 8

✦━━━•◦◆◦•━━━✦

MY NAME IS IAN MUSK

T he mobile phone rang, and its piercing ring tone broke through the later summer wind and wrapped itself around me in the cockpit of my Chrysler Sebring. Early mobile phones, "cells" as we called them, had only one simple and deafening ringtone. They could not play myriad noises, music, and grunts like modern smartphones.

It was a beautiful summer afternoon typical of late July in Northern Utah. Life in that part of the world slowed to a crawl in the summertime as the grass dried to a brilliant orange and tan hue, and the skies radiated a bright blue like very few other places. If there ever was a place on Earth deserving of the term "God's country," this was it. Late July in Logan, Utah, offers the perfect weather for a top-down drive home, and I was not going to pass this opportunity up. Friday afternoons at work were almost always deserted after lunch as smart people headed home to get a jump on the weekend or, if they were industrious Mormons as they were apt to be, they would mow the lawn and weed flower gardens before Saturday arrived. Competition for having the best-kept and orderly façade in the Mormon neighborhoods is very high, and my colleagues were fierce competitors in this respect.

On the other hand, I tended to stay a bit later in the afternoon, savoring the quiet and the opportunity to be more productive than usual. This Friday, I had nothing going on at work or home but decided to head home early, to take in some quality time laying in the hammock in the shade of the large 100-year-old Cottonwood trees that graced my 5-acre ranch property. I dropped the top down as soon as I started the car, turned on the air conditioning to take some heat off the late afternoon air, and soon found myself motoring down a deserted two-lane country road through alfalfa fields toward home.

I have always been careful not to talk on the mobile phone while driving. My brain could be better suited to simultaneously processing speech, thought, and driving. This was especially so with the top down on the car. I looked at the phone and could see that it was a 650-area code – Palo Alto, California. The ringing telephone persisted like a siren in a highway tunnel, and I tried hard to imagine who from there might be calling me on a Friday afternoon. I lived in Silicon Valley in the 1970s and 1980, but by 2001, I had retained no friendships with anyone living there and could not recall any business associates there either. At the time, and for many years later, I had no idea of the portent of this phone call to me or history.

On this Friday, I had spent the day at work arguing with what I came to call "blockheads and bureaucrats." My job running business development for a small university-owned space laboratory was generally low-stress, but the indecisive nature of my management and the staff that I often dealt with made the job less than stress-free on many occasions. I took this job in business development (otherwise known as "sales" outside of the aerospace industry) in the mid-90s after working with US Department of Defense programs in the former Soviet Union. Those projects also had their own particular "bureaucrats and blockheads," often more deadly than the general university department variety. Still, the generic problem facing you daily was much the same as in the former Soviet Union as it was now in Logan, Utah. By 2001, my problem du jour was the excessive growth of the enterprise and the attendant growth problems within the organization. Being a university-owned corporation operating NASA and DoD contracts for space hardware and research, SDL was a unique proposition.

Management was typically more focused on ensuring stability for the workforce and minimizing surprises than it was on innovation and changing the world. I was different and a fish out of water in this environment. I am a hunter that thrives on finding new opportunities, tracking them, hunting them, and turning them into reality. I thirsted for the thrill of the chase, for the challenge of a problem that defies a solution. There is no greater pleasure in life than to complete this cycle. Initially, I was at home in this new job in the mid-1990s when the organization needed to grow, and its management gave me the resources and freedom to hunt down and capture revenues and projects. However, by 2001, our revenues had grown from some 35 million dollars per year in federal contracts to over 75 million dollars yearly. I could have taken a vacation for several years, and my management would have been happier than they were with my continued hunting, which made management challenges on their side of the fence. So, my days became filled with developing strategies for convincing the management of the benefits of more growth and the importance of "being in the game" for all the new projects and programs unfolding weekly.

The phone stopped ringing, and I saw the Motorola Star Tac screen go dim. As I contemplated who this might be calling me, the phone rang again within 30 seconds. The caller on the other end of the line would not be deterred and did not leave a message. This time I picked up the phone, opened it with a flick of my right thumb, and placed it on the green answer button. I hesitated for a second while debating whether to finish listening to an outstanding version of Jimi Hendrix's "Hey Joe" or to end this incessant phone ringing. I decided to end the debate and pressed the big green button. "Hello, this is Jim," I answered in my best professional tone and, realizing that "Hey Joe" was still playing in the background, reached down to push pause on the CD player. I still heard no reply and was tempted to hang up but repeated it. "Hello, this is Jim." I now listened to a slightly hesitant voice on the other end of the line, male, maybe in his late 20s and with a faintly English accent. "Is this Jim Cantrell?" the mystery man inquired. "Yes, it is. How can I help you?" I replied with a little more than simple curiosity by now.

The conversation that transpired over the next few minutes has become symbolic of a whole new generation of entrepreneurs and bright people who have the will, the ideas, and the means to change the world as we know it. "My name is Ian Musk, and I am an internet billionaire who wants to start a private space program. I got your phone number from a mutual friend, Bob Zubrin, who said I should call you". I was unsure what to make of someone from Palo Alto with a strangely British accent and an even stranger message. Ian continued without much hesitation: "I am the founder of Zip2 and PayPal and believe that mankind has to become a multi-planetary species to survive, and I want to do something with my money to show that this is possible". While the rational part of my mind was screaming at me to hang up the phone, the adventurous side of my mind wanted to see where this might go. "Hi, Ian," I replied, "I am happy to talk to you, but I am on my way home, and I can't hear you very well with the top down on my car." I paused momentarily and started again, "Can I call you back on this number in about 15 minutes?". Given the suddenness of the call in the middle of an otherwise peaceful and mundane Friday afternoon, I was very proud of my measured and semi-coherent response. "Sure thing. I will be waiting for your call," was all I heard, followed by the infamous "click." I closed the black Motorola flip phone and pondered for a few moments what I had just heard.

In my line of business, I hear strange ideas as a matter of course. It's what happens when you go out looking for money for new projects, and you begin to understand that we all live in a massive market of ideas, and it's often hard to know ahead of time which ones will attract capital and then be successful. That's two different things, by the way, attracting capital and being successful. One does not imply the other. Ideas come in many forms and with many goals, and it takes a patient ear and an open mind to sift through these carefully.

Some ideas seem far-fetched at first blush, like when Skybox Imaging approached me to help fund a privately built constellation of imaging satellites that ordinary users could task from an internet portal. This was, in essence, giving the ordinary citizen with a credit card and a computer the same power of overhead imagery collection that many national intelligence agencies could only have hoped for some 20 years before. It was easy to see that this idea had tremendous practical appeal and could change the way humanity views itself from above, but could it be done for the money they thought was reasonable – at less than 2 million dollars per satellite? I was skeptical but helped them anyways.

Another compelling but strange idea that I was approached with involved a group of people who thought we should put away lockers of human DNA and ship them into space, hoping that alien worlds might reproduce humans upon finding the contents and the electronically encoded instructions. Sort of a strange twist on Carl Sagan and Ann Druyan's recording of the "Sounds of Earth" that was attached to the Voyager spacecraft out into the cosmos and never returned. I also spent many years working on a program that seriously thought we could float a balloon on Mars and discover water with such a device. A mission like that might "change the way we see the planet Mars and the possibility of life on that planet," we rationalized. On this scale, someone with much money wanting to do something in space to make a point was not all that crazy.

I arrived home and drove up my long gravel driveway lined by young Catalpa trees. I loved the scent of the trees when they bloomed, and they provided some lovely shade as well. My house in Utah was a modern replica of a house built by one of our founding fathers in 1798, John Jay, still in existence in Katona, New York. I had fallen in love with the home some years ago and had found plans for a modern version of the house. It was the first and last house I built, having served as the contractor, electrician, flooring installer, and painter during my after-hours. The house was a beautiful 6000 square foot mansion set in the foothills of the Rocky Mountains overlooking the valley below. It was painted the same deep shade of the late summer wheat – a golden yellow. Its construction had left my savings dry, as do all projects, and I had been consulting after hours on some other engineering projects to help finish the house and replenish the savings. Despite my complaints, one nice thing about working for a university-owned laboratory was that they not only allowed outside consulting but encouraged it under the idea that it "broadened" the experience base of its employees. This was part of the academic flavor of my then-employer, and its open atmosphere suited me just fine. It also helped me get involved in many exciting projects whose significance I would not appreciate until much later.

I headed into the house and greeted the kids. They were always happy to see me when I returned home from Work. I headed directly into my library on the home's south side. It was a large room with 20-foot-tall ceilings. It was a perfect place to discuss big ideas. I looked on my mobile phone at the 650 number and rang it back after having sat down. My curiosity was getting the best of me by now. The phone line rang the prerequisite five times before a fax machine picked up the line and sent a shrieking howl down the line and into my ear. I hung up the phone and thought to myself, an internet billionaire that only uses a fax machine for phone calls? I sat and looked quietly at the cell phone, started going over what was said in those short few minutes, and decided to do some research.

The internet was young in the summer of 2001, but the millionaires and billionaires were starting to stack up. The tech market had risen to unbelievable heights with dot-com fever, and prices paid for a simple internet presence defied the most open-minded financial analysts. Then, I could be counted among the skeptical mindset preferring the "bricks and mortar" approach to developing real company value over the internet, which had yet to show its utility to make any real cash. I grabbed my laptop computer and plugged it into the phone line to access the Earthlink dial-up internet. In a symmetric reversal of the unsuccessful return call to the mystery man, the computer now echoed a successful data connection's shrieking and ping-pong noises. I immediately entered the search engine site and typed Ian Musk, an internet pioneer.

Nothing came up. The skeptic in me was starting to take its place on my left shoulder and beginning its banter. What a crock of crap ….

Some guy calls you out of the blue, says he's rich from the internet, and wants to start a private space program. Are you that gullible? My inner skeptic started murmuring. But I wanted to believe this story. It just sounded so refreshing and unusual. And it suited that point in life where I was beginning to tire of sparring with the stability-seeking Mormons I worked for. I remembered something else that he mentioned PayPal. What in the world is that? I asked myself. The magic of the internet search engine answered that it was an online payment system that was taking on banks and allowing users to bypass credit cards, banks, and lines of authority when moving money around. I liked the idea the more that I read about it. Perhaps, I thought, there may be something to this story. I dug deeper and found that someone with a similar name had been a founder of PayPal along with Peter Thiel. They were based out of Palo Alto. This story started to make some sense.

Elon Musk was a young South African guy who came to this country with only a few dollars in his pockets and grand dreams. While he had come from some privilege in South Africa, he had been determined to arrive in the US and make his way. This is the place, after all, where people can make dreams come true, and a person with a good idea, ambition, and hard work can make their way. Elon had dropped out of Stanford University and started a company called Zip2. He had been determined to make money from the connection between traditional newspaper publishing and the growing internet user base. He set out with his brother Kimball to build the software that would connect these two worlds and made some initial deals that put the company on sound financial footing. He and his brother eventually sold Zip2 to Compaq Computers for 300 million dollars in cash. Elon took that money and reinvested it in his latest venture he called PayPal. It was as American as a story can get and quite charming at that. I could personally relate to him, his burning ambition, and his path through life, and I felt a deep connection to his ideas.

I had not managed to make it entirely through my full due diligence on Ian Musk other than determining that his real name was Elon Musk and that he had been removed as CEO of PayPal a few months earlier by the time my cell phone rang again, this time with a different number but a 650-area code still. I answered, and this time the voice was familiar. It was Elon. As is characteristic of the man, he wastes no time on pleasantries and starts right into the topic. "You didn't call me back," he forcefully stated. "I did, but all I could get was a fax machine, and I couldn't leave a message on that," I replied confidently but politely, not knowing yet what to make of this person who was still largely unknown to me. His voice lightened as he gave off a boyish short chuckle that was quite disarming. "Well, I guess that's right. Sorry about that. I am in my car now on my mobile phone and don't have very long to talk," he replied, getting right to the point. "Look, I am calling you to see if you can help me on a project that I am thinking of taking on," with that, he stopped talking. "Sure, no problem. What do you have in mind?" I replied. What was to follow has become the same speech that I have heard him give, read in the trade and popular press, and have seen him repeat on television.

"Well, you remember my point about Mankind becoming a multi-planetary species?" he said, not asking a question but making a statement again and only pausing for a short breath. "Well, I am convinced that I can do something with my money to show that this is not only possible but, moreover, that life itself can spread beyond Earth." This was an interesting point, even if it came from someone I had considered entirely insane no less than 30 minutes ago. Elon continued talking and did not miss a beat with his problem analysis. "I have been thinking about a few ideas, and it seems they all require a fairly sizable rocket to accomplish." Elon paused momentarily, I supposed to see if I was still on the line and listening. "Sure, that depends on the mission and what you are sending," I added in my most matter-of-fact and wise tone. "So, what can I do to help you with this?" I asked, still trying to understand why Elon was calling me of all people in the world. Elon's answer suddenly clarified why we came to be in the same place and time on this subject. "So, I have concluded that to accomplish this goal within the budget that I am willing to allocate, we will need to launch with Russian rockets, and Bob Zubrin tells me that you are the best expert on those in this country," without a doubt convinced that I was the guy to help him.

Bob Zubrin, the ever zany and most creative person I have ever met in the aerospace industry, was a veritable factory of ideas and energy to push those ideas forward. I love Bob and all that he stands for and fights for. Bob Zubrin is a well-known author of The Case for Mars, is an oft-cited futurist, founder of The Mars Society, and has made numerous television appearances promulgating his ideas for colonizing Mars, for affordable energy, and Libertarian causes around the world. Bob is, and always has been, a most unusual fixture of the aerospace industry and a true source of original thinking. His wit is quick, his tongue sharp, and his eyes piercing green in intensity. He is quick to analyze a problem and propose a solution, maybe any, but a solution that he is prepared to argue to its death. Descending from Russian Jewish heritage, Bob has fighting ingrained into his soul and makes his ancestors proud. He started life as a taxicab driver in New York City and describes this as one of his most remarkable experiences in understanding our country's true soul and greatness. Bob later studied Physics at NYU and earned several degrees with a Ph.D.

I first met Bob when he worked for a defense contractor Martin Marietta. I contacted him in 1991 while working for the French Space Agency on a Soviet-French Mars mission (Mars 94) and was looking to return to the US. Working for a large defense contractor, Bob was an obvious entry point for a job in what I hoped would be a US-based Mars exploration renaissance. Indeed, with such bright ideas coming from Martin Marietta, they would be front and center of the new human mission to Mars. President George Bush had just announced that the US would be "going back to Mars," said with a twang that only a politician from Texas can muster, and indeed the next significant phase of humanity would be where I wanted to be.

Bob's article in Aerospace America was entitled "Mars Direct" and described a mission design to get a human mission to Mars with the minimum government investment in space infrastructure. My time in France had estranged me from the day-to-day happenings in the US aerospace world, and Bob had created a firestorm with his "Mars Direct" concept, which had promised to land humans on Mars at 1/10th the cost of the traditional plan. It also had a folksy appeal to the public and America's pioneer roots by proposing a technical solution that would have the astronauts "living off of the land" and making fuel for the return voyage from the Martian atmosphere itself! Zubrin's scheme would place an "advance party" lander well ahead of the landing team, and this lander would be tasked with making the methane and oxygen from the Martian atmosphere to fuel the return vehicle. Zubrin's scheme had the pizzazz and practicality it needed to get extensive exposure in the NASA and larger aerospace community. His promotion of the concept was far and wide, and he published numerous articles about the idea in the early 1990s.

As I sat, phone to my ear, listening to Elon's speech about 'becoming a multi-planetary species' and thinking about how to respond to Elon's flood of information, he suddenly changed tack. He asked me a series of alarming questions. "Where are you?" he asked with no hesitation. I replied as simply as possible, "Right now, I am at home near Logan, Utah," not wanting to be too forthcoming. Something in my DNA still would not allow me to trust this compelling but random voice on the other end of the phone. Does it have an airport nearby? Elon asked to which I replied, "Yes, a small municipal airport which I can see from where I am sitting right now, "worrying about revealing too much information. Elon pressed on in his characteristic style. "I have a private plane, and I can be there in the morning," followed by a pause after which he expected a response like a chess game. At this point, my mind was declaring some alarms. 'Red flags on the field,' I screamed to myself, becoming alarmed with the prospect of someone that I barely knew showing up at my home, my sacred oasis away from the world inhabited by my children. I began to think very quickly and came up with a scheme that I am still proud of. I lied. I am not proud of lying but proud of the cleverness of the lie. It was a thing of beauty.

"Elon, I tell you what," I stated with great certainty. "I have to fly commercial out of Salt Lake City Sunday evening, so let's meet before I fly out." I paused for a response and, hearing none added, "You can fly into the executive terminal, and we can meet in the Delta Crown Room behind security in a conference room." I paused again and heard no response. So I added, "Bob Zubrin can join us." My thinking was deliberate and partially a result of my paranoia from my experience with the Russians. I reasoned that by meeting behind security (and remember, this was pre-9/11, where we could go behind security without a boarding pass), Elon could not be packing a weapon. The police were nearby in case he was completely insane. After all the Russians had put me through, I could never be too cautious with my safety.

Elon's response was straightforward and surprised me. "Excellent," Elon stated with great enthusiasm. "I will phone up my pilot and confirm that we can do this tomorrow," he continued as if this was much like getting in the car and driving down the street.

His own pilot? I wondered out loud after we hung up the phone.

As I hung up the phone, I sat quietly in my library. It was a quiet place where books, literature, and French antiques surrounded me. As I reflected on the prior hour of my life and all that I had heard, I could not shake the sense that my world had become very different in a very short period. How different it would become; I was only beginning to understand.

Later that evening, I received an email from Elon that he could meet me in Salt Lake City. We agreed to meet at 2 PM at the Delta Crown Room conference room on Sunday. I called up Bob Zubrin and explained the idea to him as well. He had already been engaged with Elon and was apparently "in the know" about this event anyways. Bob said Elon had attended a Mars Society meeting in Palo Alto some months before looking for ideas for 'exploring Mars'. Like me, few people in the initial meetings knew who Elon was and understood the extraordinary role that he would be playing in our lives. Bob had gotten wind from The Mars Society chapter members of an "internet entrepreneur" asking questions about private Mars missions. Bob attended the meeting with Elon as the Founder and President of the Mars Society.

Bob quickly brought me up to speed on Elon's mission ideas that could "demonstrate that humanity could become a multi-planetary species." The original idea that Bob and Elon had conjured up was to fly living animals, namely mice, from Earth to Mars on a free return trajectory and demonstrate that life could live and breed in the trans-Mars environment. I privately derided this idea later as "rats in space," but it had a certain charm. As I had proposed, we agreed to meet with Elon in Salt Lake City. It was easy for Bob to come from Denver, and we could discuss the mission idea.

Sunday morning came, and I headed down to the Salt Lake City airport from my home in Logan, Utah. Being the aerospace equivalent of a traveling salesman, this was a well-worn path for me but one I always enjoyed. This morning was tranquil as it was a Sunday in Utah, and all the good people attended church. On the other hand, I was heading to meet with people to use an internet entrepreneur's fortune to give mice a wild ride in space. The contrast stood out, and I enjoyed the moment's irony.

The path from Logan to Salt Lake City passed through the beautiful Wasatch Mountains, oddly reminiscent of places in Switzerland with well-kept fields and a sense of order that is hard to describe. The mountain passes, rising as high as 8000 feet through 10,000-foot peaks, soon give way to a flat, featureless plain inhabited by all but dead Salt Lake to the west. Adjacent to the lake is a very narrow stretch of land occupied by descendants of the original Mormon settlers and families that built the first railroads through this part of the world that connected the new and old worlds. As you drive further south along Interstate 10 towards the airport, the scenery gradually becomes more and more modern and distinctly less like the Swiss-like areas of Utah behind you. Parts of the Salt Lake Valley remind me of the area I grew up, 50 miles east of Los Angeles.

I arrived at the airport and parked my car in the general area of the parking lot as I always did. Being such a frequent traveler, I had developed a strong sense of routine, and it seemed as if little thought went into following this routine. That was very good this morning as my mind was far from thinking about what exits to take and where to park the car. I headed into the airport, and in a scene unlikely to be repeated anytime in my lifetime, I passed quickly through security without emptying my pockets, removing my laptop from my bag, showing a boarding pass, and showing my identification. With smiles and flowers in their breast pockets, the private security guards scanned my baggage in the X-ray system and wished me a good day. This was a little over two months before the tragedy of September 11th, 2001, and life was peaceful, and airports were much less hectic places to conduct business. I turned left from security and headed towards the D gates at the airport where the Delta Crown Room was located.

The Delta Crown room was a safe-haven refuge from the madness of the world's airports. Quiet, serene, and full of politeness. It was the perfect place to meet, and fortunately, they offered private conference rooms for rent. I rented one there and was quite pleased with it. Its large space, spartan chairs, and opaque windows perfectly symbolized the events to come. Sitting there waiting for Elon and Bob Zubrin to show up, little did I realize that the next two hours were about to utterly change my life and put us all into a unique place in history. I sat quietly and had second thoughts about even coming down and renting the conference room. "Did it send the wrong message?" I wondered to myself. After all, this guy might be a complete flake and not even show up. I knew that Bob would show up, and in any case, the two of us would have a good time reminiscing and making fun of the various characters that inhabited the Soviet-like aerospace industry.

Elon was the first to show up at the meeting. He came into the room with a powerful personal presence, looked me directly in the eye, and shook my hand firmly as he formally introduced himself. Elon was taller than I had imagined, with a six-foot-tall frame. He had a very disengaged look in his eyes, however. You could tell his mind was always processing some other subject, even though he may still be talking to you. We began our conversation with his idea of going to Mars. He wasted no time getting to business as he explained his idea about sending mice to Mars. At about this point in the conversation, Bob Zubrin entered the room. Bob is a short, intensely intelligent man with a New York accent. His personality was such that he immediately took over the presence of the room. Bob has a highly fast-moving mind with even faster-moving lips. This had the effect of stealing the energy from the room. Even Elon needed help keeping up with him. After a few minutes of standing and talking, eventually, we three sat down. Elon announced that he wanted his friend, Adeo Ressi, to join us and that we could bring him in on a telephone line.

At this point, Elon took command of the conversation and began to outline his overall goals. He outlined his mission in mind: sending mice to Mars and bringing them back to prove that they could survive the transit from Earth to Mars. This would extend the idea that humans can go to Mars. That was his goal, and it was clear from this conversation that is where his mind was. He also explained that he could not afford a US rocket and that we would need a Russian rocket. That is why I was there. The entire conversation was very rapid and very efficient. We decided that we needed a small crew to design the mission so that we could size the rocket.

I was responsible for leading the effort and putting the design crew together. Elon agreed to pay for the effort. I had people in mind that I had worked with before and would start recruiting just as soon as we were done with this meeting. After about an hour into the conversation, Elon put his phone down on the table and called his friend Adeo. Adeo was a friend from college and was intent on helping Elon with this crazy mission he wanted to accomplish. Adeo was a different sort of person than Elon. He was a pure entrepreneur like Elon but much more pragmatic and had his "feet on the ground." We briefed Adeo on our plan, and I gave him some background on myself and the team I would put together. Adeo agreed that it was a good start.

We ended the phone call and the meeting after about two hours. Elon, Bob, and I headed into Salt Lake City for dinner. We ended up at my favorite place, the Market St. Grill, and had a nice dinner. We spoke mainly of our Mars mission. Bob strongly advocated sending mice to Mars, but I was more skeptical. We ended the visit by returning Elon to the airport, and I returned home. It was clear I had much work to put a team together to define this mission and then acquire the pieces on Elon's behalf.

The following day I called my colleague John Garvey who I had worked with previously on several Russian programs, including a Russian rover. Like me, John existed somewhere outside the normal bounds of corporate America. He had worked for McDonald Douglas but found disfavor in corporate America and had begun a consulting practice. I called John to see if he was interested in helping with this program, and he was. He also had some names of people who could help. We recruited John's friend Chris Thompson who would later become a fixture at SpaceX and went on to play lead roles at Virgin Galactic, Astra, and now my rocket company Phantom Space. I also called several other colleagues, and we recruited an all-star team. One major recruit was Mike Griffin, who later became NASA administrator under Bush and Secretary of Defense under Trump. Additionally, we added some key individuals from the JPL Mars program who agreed to work on the side with us.

At the time, I considered John Garvey a good friend and trustworthy. John later came to disappoint me greatly. I eventually found him paranoid, egotistical, and unable to work in a team environment without sabotaging the entire enterprise. In retrospect,

I am happy that he turned down a role in SpaceX, as this would have been an unnecessary disruption in that enterprise. I became business partners with John nearly 15 years later, and he single-handedly destroyed the very successful business he and I spent years building with his paranoia and scheming.

Chris Thompson and I first traveled to Bob Zubrin's shop in Denver. We discussed his idea for sending mice to Mars. This wasn't Elon's idea but rather Bob's. He saw it to demonstrate that humans could survive a trip to and from Mars. Elon had the goal that mission was to inspire humanity that someday they would be able to travel to and live on Mars. Chris and I were not convinced that this was the right approach for the mission and began to think about other ideas that might result in the same outcome.

I consulted my colleagues at Paragon in Tucson, Arizona. Jane Poynter and Taber McCallum ran Paragon. During its initial two-year mission, this husband-and-wife pair had been one of the original 'Bionauts' living on the inside of the biosphere just north of Tucson. They were unusual people and had unique ideas. I knew them through some professional organizations I belonged to, and they had an exciting vision for putting a greenhouse on Mars. They had proposed this greenhouse concept to NASA, who were too conservative to consider such an experiment. In any case, the idea of growing a plant on Mars consuming Martian CO_2 and producing oxygen from the plant through photosynthesis symbolized the concept of life on Mars much more clearly. Chris and I reported back to Elon about this idea, and he liked it a lot. As we began calling the mission, Mars Oasis officially became a greenhouse experiment rather than the 'mice to Mars' mission. This shift upset Bob Zubrin. He often tried to convince me that the greenhouse idea was "dumb" and that "astronaut mice" were better. This was not my decision, however.

Mars Oasis lander concept with plant growth chamber shown on top. Image Credit Lavotchkin

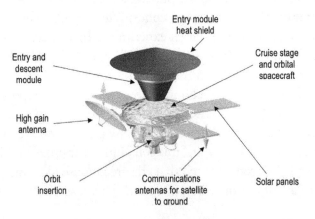

Mars Oasis spacecraft concept. Image Credit Lavotchkin

Bob eventually became separated from the Mars Oasis program. Now that we had chosen a baseline payload and Paragon would provide it, we had begun to think about what kind of lander we might put together for this mission. We had two options. Option one was where we recruited an American team that had done the Mars Pathfinder mission for JPL and NASA and build the lander in the US and launch it from Russia. We had personal connections with this group, and they were very interested in doing something like this. The second option would be to pay a Russian group like Lavotchkin to build it and fly it out of Russia. We would provide the payload.

Chris began working on the design for the American lander. He also worked with Paragon on the design of their payload. I took the lead on involving the Russians and putting their plan together. It just so happened that several of the senior leadership from Lavotchkin were in the United States late that year to talk to The Planetary Society. We arranged to meet with them, and they agreed to do a study for us. We gave them the specifications for the payload, and they asked for $10,000 in cash to conduct the study. Seeing somebody like Elon struggling to find $10,000 in cash was funny. Of course, he had the money, but getting it in cash was a completely different story. We could not get it out of ATMs because of the limits there, and his bank accounts were in very nontraditional Banks, which were unavailable in the Pasadena, California area. We finally managed to produce $10,000 and hand it to the Russians. It felt more like a drug deal than something that might lead to a mission to Mars.

The team spent the rest of the summer concentrating on our US Mars Lander design and examining the Russian rockets that might take the lander there. As we worked on the system-level design of the lander, we began to understand its' weight parameters and some of its requirements for entry at Mars. I engaged Jim French, an old colleague and mentor, to help with the Mars trajectories. I also engaged the members of the JPL Pathfinder group to help us with the lander design.

I led the effort to choose the Russian rockets and narrowed it down to two. The first was the Dnepr. The Dnepr was based on the R-36 MUTTH Intercontinental ballistic missile (ICBM), the SS-18 Satan, by NATO. It was designed in the 1970s by the Yuzhnoe Design Bureau in Dnepropetrovsk, Ukraine. On April 21, 1999, the first launch successfully placed a 350 kg satellite into a 650 km circular Low Earth Orbit. Originally these rockets carried several megaton-class nuclear warheads designed as city busters. They had been converted to satellite launchers during my time in Russia, as many of these missiles had been taken out of service by treaty.

The second option was the Strela. The Strela was derived from the Soviet UR-100NU missile, also known as the SS-11 "Sego" by NATO. The SS-11 was an intercontinental ballistic missile (ICBM) developed and deployed by the Soviet Union from 1966 to 1996. The Strela conducted its maiden test launch on December 5, 2003, carried its first satellite on 27 June 2013, and a second on 19 December 2014. This vehicle was built by NPO Mashinostroyeniya, a rocket design bureau based in Reutov, Russia. During the Cold War, it was responsible for several major weapons systems, including the UR-100N Intercontinental ballistic missile and the military Almaz space station program.

By the fall of 2001, we had made quite a bit of advancement in our design. We decided to visit Moscow to meet with Lavotchkin and some of the launch vehicle providers. As it turned out, we could not meet with launch vehicle providers this early, but we did spend time with Lavotchkin to understand their offering for a design. During this visit, our friends at Lavotchkin also arranged for us to have tours of Energia, other Russian providers, and several museums of space hardware. Lavotchkin proposed a very interesting lander based on several Mars Landers they built in the 1960s and 1970s. However, none succeeded, and we were mindful of this outcome. The price was right, but Elon was unwilling to "throw away his money in some warehouse in Moscow" and thus unwilling to engage Lavotchkin.

Adeo Ressi, Elon's friend from college, was with us on this trip. In the evenings. these two often had different ideas of how to spend time than I did. One evening they asked me if I knew where the "Hungry Duck" was. I was surprised that Elon even knew of the Hungry Duck. It had an infamous reputation in Moscow. His cousin from South Africa knew of it and recommended they go there. I was holding security clearances at the time, and I had to decline the invitation to go with them. The Hungry Duck was well known to have FSB (new KGB) agents lurking in the crowd in the female form. These people were there to attract people like me and 'influential Western businessmen' like Elon and Adeo. I agreed to take Elon and

Adeo to the Hungry Duck, leave them there, and find their way back to the hotel later that night.

The Hungry Duck was a wild place where anything goes, and Adeo and Elon had a lot of fun there. When they decided to leave at three in the morning, they did not take my advice to take their passports. On their way home, their taxi was pulled over by a Moscow policeman standing on street corners in the morning. No harm was had as the taxi driver negotiated a release for them for $800 each.

We were to meet a van to take us to a museum at 10 o'clock the following day. I was in the hotel eating breakfast downstairs, waiting for Elon and Adeo. Both showed up very late. Adeo showed up first and only wanted coffee and didn't say much. I asked him how the Hungry Duck was, and he just looked at me with bloodshot eyes and rolled them back into his head.

Elon showed up next and looked much the same. I asked him what happened at the Hungry Duck, and he just smiled and laughed. I always admired Elon's ability to laugh at absurdities. It always seemed like a healthy way to deal with the absurdities of life. Elon finally looked at me and said, "You were right. We should've taken our passports with us last night". He then relayed the experience of getting a shakedown from the Moscow police. My response was simple and to the point: "Welcome to Russia."

Our first visit to Moscow ended with no drama, and we returned to the United States. I was pleased to return as it was early September, and the weather was getting very friendly. However, we were unaware of what awaited us on the 11th day of September as we got off the planes in New York just a few days before those tumultuous events. I remember looking up and seeing the World Trade Centers for the last time I would ever see them. I headed back home to Utah. Adeo went to his home in New York City, right next to the Trade Center towers, and Elon headed back to Palo Alto in California.

The morning of September 11 was a shock to all of us. I was supposed to head to Pasadena for some meetings with JPL, but as I got up in the morning before I caught my plane, I saw the horror of these airplanes flying into the twin towers, I immediately knew that something really, really bad was afoot. I also knew that Adeo lived on Church Street right next to the towers and immediately became worried about him. I called Elon later that morning and asked him if he heard from Adeo, and he had not.

Elon was very philosophical about the whole terror attack. He commented to me, "It's a shame. This used to be a great country, but the only choice we will have to fight against this kind of terrorism is the clampdown on individual liberties". At the time, I thought this was a bit of an exaggeration on his part. However, the time passing has shown that he was correct, given the Patriot Act passed in the US and all the abuse of civil liberties that the Patriot Act justified. We would never be in the same country after 9/11.

Eventually, Adeo turned up again at his home in the Hamptons. He sent an email to all of us letting us know that he and his girlfriend narrowly escaped death as the second tower collapsed in front of them. They hid behind a bridge over the West Side Highway and somehow survived while all those around them died from the flash of debris from the collapse. This experience permanently changed Adeo, and I sensed it. He withdrew from any more involvement in Mars Oasis as he had to think about what his life meant and what he wanted to do with it.

Elon, on the other hand, was still undeterred. We continued to work on Mars Oasis, and it was beginning to gain some actual definition. We had several review meetings where we invited outsiders to review the project and give us their opinions. Some involved my colleagues, including Chris McCormick, an entrepreneur I've known for years. Others included James Cameron, the famed Director, who, in one meeting, gave us advice on how to engage the public and media with Mars Oasis. It was always fascinating to see who would show up to these meetings. Mike Griffin also began attending very regularly and contributing on a technical level that was well beyond the technical insight of most of us.

We flew to Moscow in early November 2001. A war was underway in Afghanistan by this point, yet it seemed a world away. Mike Griffin joined Elon and me on this trip, and Mike was understandably nervous about going to Moscow. He was also holding security clearances like me and was warned of all the dangers in Moscow if you weren't careful. Nonetheless, it helped that the two of us were cleared and could vouch for the other in case our behaviors were questioned later.

Our first meeting was with Mashinostroyeniya. It was a fascinating meeting. Like most Soviet-era aerospace concerns, it was surrounded by very tall walls that amounted to about 20 feet of concrete walls covered with barbed wire on the top. Mashinostroyeniya looked like a military compound with huge steel gates and an armed guard at the opening. A professional driver took us there and explained that we had an appointment to meet with their Chief Designer. The old Soviet system organized their industrial entities around a Chief Designer rather than a CEO. No such thing as a CEO existed in the Soviet Union. The Chief Designer is the closest thing to what we see today as a Startup CEO who is a technical leader in the field and the organization's overall business leader.

NPO Mashinostroyeniya. Image Source Yandex.com

We were eventually let through the big steel gates and arrived at the front door of the building, which ironically had another set of massive steel doors. It was very snowy this morning, as it is often in the wintertime in Moscow. We exited the car and approached the large steel doors knocking on them. There was no high-tech doorbell and nothing else to communicate that you were out there except a firm knock on the door. We waited for about five minutes and eventually could hear a series of locks being undone inside the Doors.

Eventually, the doors began to open very slowly. I half expected the character Lurch from American TV, but instead, we had a Russian security guard with a fur hat and a coat inside the building. This could only mean one thing. The hallways were not heated, and the temperatures inside the building were almost the same as those outside. As would be expected, it was frigid inside.

Mike and Elon could have been more impressed. They were still dead silent as we entered the building. I did most of the talking in Russian. Elon and Mike were looking around, taking in the scene. The scene of this almost post-apocalyptic weapons factory wasn't as foreign to me as it was to Elon and Mike, having spent six years in Russia. However, entering this large fortress was very intimidating, no matter how you looked at it. As we walked down the hallway, no lights were on, and it was very dark. The corridor was very long and very wide. Elon noted that there was vinyl-covered padding on some of the doors and wondered what it meant. He leaned over to me and said, "This looks like an insane asylum," in response to the padded doors. I explained that that was generally a sign of somebody important inside, and while I didn't understand why the doors had such significant meaning, they did, nonetheless. We continued down the corridor with Mike saying absolutely nothing. We eventually arrived at the principal office of the Chief Designer.

In the Soviet days, the Chief Designer was the equivalent of a Western CEO. Usually, somebody with excellent technical capability would lead the organization's new project ideas and development projects. Since this wasn't a capitalist society, including a business sense was not necessary for the Chief Designer. The Motherland would provide funds for those projects deemed essential to the Soviet Union. The structure that survived through the 1991 coup d'état was still intact when we negotiated with them in 2001. Historically, the original Chief Designer for Mashinostroyeniya was Chelomei, the great rival to the well-known Soviet Chief Designer Korolev. We were stepping into the great history of the Soviet space program.

We were led into a long conference room with a big table with a spread of food. This was a classic Russian setup and was getting very close to lunch by now. The long table had many finger foods, as I called it, which were pieces of bread with unidentified meat on top of them and cucumber sandwiches. This was a prevalent Russian staple.

Our Russian hosts joined us almost immediately after we entered the room, and we began shaking hands and introducing ourselves to each other. The Chief Designer was a very nervous-looking gentleman with teeth that were missing and clothes that were proper but ill-fitting. He looked as if he had lost much weight and had not bothered to buy clothes that matched his new weight. He was a very serious-looking man and had severe questions for us. Nonetheless, we started the meeting, a tradition in Russia, with a vodka toast on an empty stomach. Perhaps it was a ploy by the Russians to make us softer, but it was consistent with my experience in Russia. Once we toasted each other and our respective countries, we began eating the finger sandwiches on the table and interacting socially with the Russian team. Many wanted to know where we came from and what we did. Mike was very quiet and talked very little. Conversely, Elon was effusive and spoke about his background in South Africa, PayPal, Zip2, and all the Internet things he had done. Strangely, the Russians had not heard of anything Elon had done and considered him a braggadocio.

After lunch, we briefed the chief designer on what we wanted to use the rocket for. We began the conversation by discussing Elon's idea of making humanity a multi-planetary species. It's important to realize that Elon was, at the time, a twenty-something "kid" from Silicon Valley who had done very well financially and dressed much like the other entrepreneurs in Silicon Valley. He dressed poorly in the eyes of the Russians, which was something of great import to them. I had warned Elon about this and how the Russians would judge him based on his appearance. Nonetheless, Elon dressed how he pleased, impacting the Russians' view of him. The Russians seemed to think I was the boss, but I would tell them in my pigeon Russian that Elon was the man with the money and made the decisions.

As we explained this to the chief designer, I could tell he was agitated. As Elon continued to talk about the idea of humans living on Mars to escape a dying planet Earth, he became visibly angry, and he spit on our shoes. He first spit on Elon's shoes and then on my shoes. Following this, the Chief Designer stood back, not saying a word as he looked at our faces for our reactions. Elon turned to me and said, "I think he spit on us." I didn't know how to respond. I finally responded, "Yes, I think that was intended to be a sign of disrespect." Elon's stated that the meeting was over and that he didn't like these guys.

I explained through the translator that we appreciated their time and would get back to them. However, from the behavior of the chief designer, it was clear that he did not want to sell us one of his rockets. I understood his conversation with one of his colleagues while we were leaving; he saw this as a very unrealistic program and didn't want anything to do with it. He continued in Russian, stating that what he built at Mashinostroyeniya was a serious weapon of war, and he only wanted to deal with serious people. We left shortly after this diatribe, thanking them for their time. By this time, we had spent most of the day and returned to the hotel.

I continued working into the night, but Mike and I had dinner together that evening. We discussed what he had seen and, that this was his first experience in Russia, how overwhelming it was to Mike. We also had the various prostitutes sitting down at our tables in the hotel during dinner to ask if we would buy them a drink. Mike and I had been trying to spot this behavior, but this overt action was obvious. We bid them farewell, finished our dinners, and returned to our hotel rooms.

The next day we went to Kosmotras, located in the center of Moscow, in an old home. These people were a bit more westernized in their business practices as they had been working with Morton Thiokol, who was building the US space shuttle boosters at the time. To demilitarize the SS-19 for compliance with the Start II treaty, the Soviets dumped the toxic fuel out of these rockets into the soil and stored the boosters filled with nitrogen for preservation. During the 1990s, the US offered Russia a program where Morton Thiokol would work with them to convert the toxic fuels from the SS19 into benign fertilizers. This was undertaken with US funding and personnel from Thiokol, who worked with the Russians to teach them how to turn nitrogen tetroxide and hydrazine into fertilizers.

As we approached the bright yellow building in the snow, it felt friendlier than our last encounter at Mashinostroyeniya. I was more optimistic that we could put a deal together here, and of the two rockets, this was the more interesting one. We had a flight later, about 2:30 PM, from Shermatayevo airport to New York, so we began earlier this day. It was about 9 o'clock when we arrived, and we were greeted by several friendly Russians who offered us coffee immediately. All three of us were happy to have excellent warm coffee on such a cold day. We went into a conference room where we were greeted by four Russians, three of whom I already knew. The fourth one I didn't know, and I called the "Blue-Eyed Colonel." He was the decision-maker despite being the youngest of the four.

We began the conversations with the usual social chitchat, and Elon gradually began to explain his idea of making humanity a multi-planetary species. The Russians seemed to tolerate this discussion much better than the Mashinostroyeniya Chief Designer we had just encountered. I could tell, however, that there was some skepticism still. After all, Elon was still a twenty-something poorly dressed Silicon Valley entrepreneur asking for a serious business deal. Being poorly dressed was always a sign of disrespect to the Russians, and they would mention this between them, thinking that none of us understood Russian. I didn't let on that I understood Russian at all. Our meetings were held in English, and we purposefully kept it that way.

Elon explained that he wanted to buy two rockets and was willing to pay $4 million each for them. Kosmotras said they would sell them to us for $8 million each. Elon was quite incensed by this price gouging as we knew that several other companies had recently bought them for $2 million each. Elon had decided to be generous in offering four million each. After some negotiation, where the Russians weren't willing to move, we decided to leave. The Blue-Eyed Colonel was not impressed with Elon, and I heard him call him a "little boy" in Russian. This was the ultimate insult in Russian society, and I knew the negotiations were over after hearing this.

We gathered up our luggage and headed off into the December Moscow snow as we headed towards the airport. We were silent during our taxi drive to the airport. As we arrived at the airport, we passed through the usual lines of security at the airport, which takes hours, and finally arrived at the boarding area for our flight. It was a Delta flight straight to New York, and it always felt like being on American soil again once we boarded. This time was no different.

The plane eventually taxied down onto the runway and took off. After about 10 minutes, the stewardess returned and asked if we wanted anything to drink. Mike and I ordered a whiskey and wanted to celebrate being out of the evil Soviet Union. Elon was busy working on his computer several rows ahead of us. He did not order any alcohol or anything to drink. He was too busy. Mike and I sat there drinking our whiskey, and its effects began to be felt. Mike nudged me with his elbow and said, loud enough for Elon to hear, "What the hell do you think that idiot savant is doing up there clearly.?" My response to him was, "Probably making plan nine to save the Earth." Mike and I laughed at our jokes. Elon turned around and said, "Hey guys, I think we can build the rocket ourselves." Mike and I looked at each other, resisting laughter.

I thought for a few seconds about a response. Finally, after a long pause, I said, "Elon, there are 100 dead bodies in front of you, and you're going to have to walk over them to get to the point of doing this. "Continuing, I said, "This is not an original idea, and many have failed already.". Elon responded to me that he had a spreadsheet of the rocket and that he wanted us to look at it. Mike turned to me, saying, "Oh, but Elon has a spreadsheet!" to which I replied, "Yes that'll make all the difference in the world".

Elon handed his notebook computer back to us, accompanied by an insult whispered under his breath, and we began to take a serious look at it. What initially seemed like a joke started to become more and more real the more we delved into it. This was a very detailed and serious spreadsheet with masses, performance numbers, calculations of staging ratios, and critical design elements. The vehicle was about the size of what became the Falcon One. Mike and I disagreed with some of the masses of the inter-stages and other parts, but by and large, this was a very realistic spreadsheet estimate of a small rocket that needed to be more coincidentally large to send our lander to Mars. It was large enough to send small satellites into orbit.

I returned the computer to Elon and asked him where he got the spreadsheet. He mentioned that he had been studying rocket propulsion and rocket design since we started working together. I knew he had borrowed some of my books from college on propulsion, orbital mechanics, and launch vehicle design. He had also been working with some of John Garvey's friends, mainly Tom Mueller and Chris Thompson, who would go out into the Mojave Desert and launch rather large amateur rockets. By large, I mean 20 feet tall and capable of going up several miles into the atmosphere, Elon was so inspired by this that he began to ask many questions. He would hang out on the weekends with this group and learn more about rockets from them, including how to design them. One of the key people he hung around with was Tom Mueller. Tom later became the Vice President of propulsion and a pivotal contributor to SpaceX. I think it's safe to say that SpaceX would not exist without Tom Mueller. At the time, Tom was employed by TRW, working in propulsion. He was also building his liquid rocket engines in his garage on the weekends. Tom is the kind of person who had a machine shop in his garage, and he had built a 10,000-pound thrust liquid engine. That impressed Elon. Elon reasoned that if guys like this with no money could put together such great propulsion systems, imagine what they could do with enough money of the type that Elon could raise. This is how SpaceX was born.

After we arrived back in the United States, there was much to do to prepare for what Elon started to call "our little company." As we did for the Mars mission, we began to put together meetings in conference rooms and board rooms of hotels surrounding the Los Angeles airport. Instead of drawing on the Mars experience, we began to draw on the collective experience of people in the US who built rockets. Some involved our core Mars team, and we relied on people we knew and their extended contacts. Our last hotel meeting occurred at the Renaissance Hotel near LAX on the 15th of December. As it turned out, most of the original core team from Mars Oasis was in attendance. It was a strange time in history as halfway across the world, the US military had cornered Osama Bin Laden in Tora Bora and lined up his end game that weekend. We did our business in our small conference room near the LA Airport, planning a revolution in the domestic and international launch vehicle competition while the rest of the world was gearing up for decades of war.

We began to interview people who might become part of the core team. As it ended up, the core team started with Chris Thompson as Vice President of structures, Tom Mueller as Vice President of propulsion, Elon as CEO, me as Vice President of Business Development, and Hans Konegsman as Vice President of guidance navigation and control.

Elon asked Michael Griffin to join us as well. Mike and I had long conversations about whether it was wise for him to join a start-up rocket company. In our earlier sessions, Mike and I recommended that we locate the company in the Ventura, California, area. That would allow us access to the deep-water port at Port Hueneme and ship larger rockets worldwide to the Cape or up to Vandenberg. Instead, Elon decided unilaterally to place the company in El Segundo, California. His logic was simple. He reasoned that most rocket talent in the United States was concentrated in Orange County, just south of Los Angeles. He also reasoned that his wife Justine wanted to be an author and that they should live in Belair to help her career as an author. El Segundo is a logical middle point between those places where Elon decided to place SpaceX. I escaped that part of the world in the late 1970s and didn't want to return. I would commute to El Segundo during the week and return to Utah on the weekends. On the other hand, Mike saw this move as less than serious on Elon's part and decided not to join SpaceX.

SpaceX facility on Grand Ave in El Segundo.
Image Credit SpaceX

We looked through empty warehouses in El Segundo and the surrounding areas for several days. Some of them were frightening. Others were too big and too expensive. We finally settled on a small warehouse on Grand Avenue, not far from the airport. It also had the virtue of being down the street from a greasy spoon café for the best breakfast possible. Our new place on Grand Avenue used to be an import-export warehouse. It was sufficient for making small rockets as planned but clearly would not suffice for a much larger enterprise. This became our new home.

Space Exploration Technologies

James Cantrell
VP of Business Development

1310 East Grand Ave.
El Segundo, CA 90245
310 414 6555 tel.
310 414 6552 fax
www.spacex.net
jim@spacex.net

Author's SpaceX business card.
Image Credit Jim Cantrell

One of my first jobs was related to something other than a new business. It was more related to helping with anything that needed to be helped with. Elon asked me to think about where we might set up an engine test facility. As it turned out, not surprisingly, we had to build our engines at this point. There were engine producers at the time, such as Aerojet and Rocketdyne. However, they were beholden to the likes of Lockheed and Boeing. Also, they were costly because they built their engines for the military-industrial complex and reflected its internal cost structure. We had to build our own engines to control our destiny and break the cost paradigm of the American launch vehicle.

My mind naturally started thinking about places where the environmental issues would be minimal to placing a rocket test facility. Several items concerned me from an ecological point of view. One issue was noise, and the second was the soil's chemical contamination. You also need the correct geometry for placing huge test stands over the side of a cliff or something of that nature to minimize the infrastructure you might have to put in terms of flame troughs and water deluge. I began by looking around the area of Morton Thiokol near Promontory, Utah. They had a substantial test capability in this remote area of Northern Utah for the large space shuttle boosters and military weapons systems such as the Minuteman ICBM.

I took my four-wheel-drive truck and headed out onto back roads trying to find some suitable farmland that might be an excellent place to place a test stand in this vicinity. I looked on maps to understand the local geometry and terrain. I spent most of a day driving around the hills near Promontory, Utah, looking for the correct ridge and placement. If it were private land, I could arrange to buy what I needed from the farmers. However, I did not receive an amiable greeting from the local farmers, and several chased me off their land. One even fired a shotgun in my general direction. The other problem with Promontory Utah was that it was too cold in the winter, and we risked being unable to test during the large part of the year.

I began thinking about other companies and where they tested rocket engines. Many of them were in California and grandfathered in because they had been there for 30 or 40 years. One launch manufacturer that had yet to make it was Beal Aerospace. Andrew Beal was a banker from Texas who decided essentially to do what Elon was trying to do five years earlier. Beal had designed an enormous rocket fueled by hydrogen peroxide and kerosene. They had managed to get to the point of some significant stage-level testing. By the time they reached this point, however, Andrew Beal had spent nearly $100 million of his own money and decided to pull the plug on the development before it would have bankrupted him. This was one of the dead bodies that Mike and I referred to when Elon told us that he wanted to build the rocket himself.

I remembered that Beal had a test site somewhere in Texas. I consulted the news archives and found it was in McGregor, Texas. I called the city of McGregor and asked them if they had an old test site left over from the Beal days. When I called, I reached the secretary and asked her the question. She told me in her long Texas drawl, "Of course!". I mentioned that I represented a rocket company that might be interested in acquiring the test site. The secretary mentioned that the city manager was standing right next to her and that he could talk to me about such a thing. When he picked up the phone, I was greeted by a familiar friendly Texas drawl. Asking about the Beal test site, the city manager said, "Well, yes, the city of McGregor owns it, and we would be more than happy to sell it."

One of the surprising things for me that Elon did when we first started the company was to buy a private jet. It was a Falcon of a nice size to accommodate a relatively large crew. We took the Falcon to McGregor and flew right into nearby Waco. We managed to be greeted by the city manager when we landed and were taken to the test site. It was a perfect site. Beal had begun to construct a vast vertical test stand for the testing of the first stage. I estimated its height at about 100 feet. It had never been completed, which was perfect for us. There had been some peroxide contamination of the soil, and the city manager mentioned that we would have to be careful if we decided to do any construction to remediate the peroxide contamination. None of this concerned us greatly, as finding a site like this was a miracle. This site eventually became the McGregor test site for SpaceX and is now a core capability around which SpaceX not only developed engines but also developed the now-famous reusability of the first stage. They performed multiple grasshopper flights in the early stages to prove their ability to land stably. This was pure heaven for many rocket engineers, and Texas was a great place to have it as both the locals and the state government were very friendly to business.

I spent much of my time traveling to El Segundo during the week and returning home to Utah on the weekends. I did not enjoy the travel or being in El Segundo. However, this was an exciting job, and I enjoyed doing it. However, I needed more faith that SpaceX could eventually be successful, and in the back of my mind, I often questioned what I was doing there anyways. My colleagues, such as Chris Thompson, did not let on if they felt the same way. We mostly talked about how to solve problems and what was next to deal with. I spearheaded the contract with Barber Nichols, who was building our turbo pumps for the engine. They were in Colorado, and I was put in charge of negotiating and managing the contract.

On our first meeting with Barber Nichols, I also picked up a project I had purchased some months earlier. It was a Maserati Ghibli that I had bought that needed to be restored. The car had burned, but the present owner had cleaned it up and found most of its parts, which were in boxes. I bought an enclosed trailer to haul the car and its parts back to Utah. I did this before the meeting with Barber Nichols and headed to the meeting with a trailer full of an old, burned-out Maserati and parts. After the meeting, Elon followed me out to my car, talking to me about the contract. He saw the trailer and asked me what was in it. I explained it was a 1967 Maserati Ghibli. Elon was a car guy and knew what this was. He asked to see it. I opened the door, and immediately we were greeted by the stench of a burned-out smell. Elon laughed and asked why I would buy an old, burned-out Maserati. My response was in jest that I purchased this so that we would have something to launch on our first rocket attempt into space.

Elon thought this was hysterically funny and started mimicking my joke. He asked us, "Why did you start a rocket company, Elon? Answer: so that we can launch old burned-out Maserati's into space." This was typical of Elon's humor, and I was happy to have at least gotten a positive response from him. When he later launched his Tesla into space on the Falcon Heavy, I wondered if this moment had inspired that idea.

The rest of the summer of 2002 was uneventful at SpaceX. I still could not understand how we would build a rocket in this warehouse where Elon refused to paint the floors, and we had insulation falling from the ceiling. It seemed like a bridge very far away from me. I did calculations in my mind and knew that this rocket would require at least $100 million to complete. Elon had told me he would put $100 million of his net worth into SpaceX. He said that if that were not enough, we would shut the program down. Based on that, there is no way to raise money to build the rest of the rocket. I was very wrong on this account.

My doubts began to nag at me, and I started questioning what I was doing. I knew that Elon aimed to send humans to Mars and eventually build a settlement there. Strangely, this was not something that I was very passionate about. Maybe it's because I didn't consider it a realistic kind of thing to be doing. Or perhaps I just preferred something a little more modest. Either way, tensions began to build up between Elon and me.

The tension peaked when he asked me to price what I thought the Falcon One tanks should cost. I had done all the cost analyses on the Falcon One and the Falcon Five and sized them for the prevailing satellite launch market. And now he was asking me to estimate further some of the more expensive pieces of the launch vehicle. I put pen to paper and devised an estimate for the first ship set of tanks to be under $1 million. This number is admittedly high for a recurring cost, but it considered all the tooling and nonrecurring engineering that would have to go into such a design which would reflect in the first unit cost.

Elon received my email on this pricing, and he became furious. He called me on my cell phone. I was arriving at the Salt Lake City international airport. I was driving around to find parking in the garage, and he began yelling at me about the tanks. The garage was very crowded, and I had no luck finding a place. I was also late for my plane. Elon yelled at me that if this set of tanks cost this much, he would "Eat his ball cap." While trying to find a parking spot and dealing with Elon's tantrum, I ran over a piece of shrapnel and shredded my front tire. Fortunately, I found a parking spot for the car finally and parked it there. I finished with Elon, and we agreed to research how tanks were made together. I did make my flight to SpaceX, where Elon and I had a long discussion about rocket tanks.

As a result of our disagreement over tanks, Elon and I decided to visit tank manufacturers ourselves. For this, Elon flew to Salt Lake City, where I had arranged to meet with several industrial tank manufacturers. We also looked at how tanks were made to an industrial standard. Most of the tanks we saw were made from rolled sheets of steel or aluminum and were welded along the edges of the sheets. This is likely the most inexpensive way to make the tanks. In the end, this is the method that we chose to make the Falcon One tanks and if you examine the Falcon Nine tanks today, they are simply a more sophisticated version of this type of construction. In the case of the Falcon Nine tanks, they are made from aluminum lithium alloys and are friction stir welded instead of TIG welded along the edges.

The brutal treatment I received over the tank cost still did not sit well with me. I did not enjoy being yelled at, and Elon was prone to yelling. This, on top of my doubts about the success of SpaceX and my not being aligned with Elon's objectives for SpaceX, led me to start thinking about leaving. I had many people offering me jobs at the time. Another thing that most people don't realize about Elon in the early days is that many in the industry saw him as a charlatan and a liar. I certainly didn't see him that way. Still, many outside SpaceX were unwilling to believe his grandiose boasts and thought him to be just one more person who came into the aerospace industry seeking more attention than a desire to do something. His external credibility was near zero. I made many introductions for him to senators, Pentagon leadership, and NASA leaders. In essence, I felt like I was spending my credibility and giving it to Elon.

It was late August 2002 when I decided to leave SpaceX. My reasoning was quite simple. I decided I could make more money as a consultant and would best serve Elon by leaving early rather than later in the company. I wrote my resignation letter to him; surprisingly, he didn't seem too bothered. Maybe he expected it. Perhaps he didn't care. Either way, I was done at the end of September and ready to move on to a new life. He asked me to help find someone who could replace me. I suggested Gwynne Shotwell. I have known Gwynne since she came to the aerospace industry from the automotive industry some 15 years prior. I always liked her. She was very competent and good at sales. I always saw my position with Elon as ultimately being sales. However, you did what you needed to do in a start-up, and I was more involved in administrative and technical matters than in any new business. They interviewed Gwynne at my request, and she was very well-liked. It's interesting to note that since this time, Gwynne has become a very trusted number two to Elon and essentially runs SpaceX. I sometimes look back on my decision with dismay, thinking about how much money I walked away from. I had a lot of SpaceX shares that had yet to be vested. But I also saw my decision to walk away as making room for somebody like Gwynne, who served Elon very well. I remain very philosophical about it to this day.

It was the right thing to do for me. I always tell people that I would have to be a different person to have survived that environment with Elon. That is a simple fact, and I still believe it very much. It's even more believable when I talk to some of the early people I worked with that left after ten years. They had a hard time making it the entire ten years. Some of them ended up otherwise perfectly healthy. Others suffered heart attacks from all the stress. It was a challenging place to work. I admire anybody who was able to survive it.

CHAPTER 9

────────◆────────

BUSINESS AS USUAL

*"Auto racing, bullfighting, and mountain climbing are
the only real sports... all the others are games"*
– Ernest Hemingway

F ollowing my exit from SpaceX in late 2002, I started a consulting
practice named StratSpace that later grew into a very successful
business. StratSpace initially focused on military applications
of small satellites. With the post-9/11 geopolitical changes as a
backdrop, military interest in small satellites (once considered small
toys and the sole domain of academics) began to have fundamental
roles in defending and eventually replacing much more expensive
and massive military space assets. The technology developed by
universities and small commercial companies started mixing with
the more significant financial resources of the military and Venture
Capitalists to produce the technological core of New Space. In what
seemed to be a repeat of my Soviet experience, I was part of a group
populated by strange bedfellows: New Space entrepreneurs, the
US military, and the intelligence agencies. This period paved the
technological and financial path for the New Space industry and
involved many interesting individuals and instructive stories linking
them together.

Many early technological and financial connections existed
between the military and New Space. These two disparate worlds
worked together at arm's length to simultaneously result in a new

industry and branch of the US military: the Space Force. Many of the same people involved in the early internet money-fueled space endeavors of the '90s and early 2000s joined forces with a new generation of younger entrepreneurs to form companies such as Skybox Imaging (sold to Google for half a billion dollars), Planet Labs, and Virgin Galactic. My consulting firm and many of my early SpaceX colleagues played a central role in this emerging set of markets and gave me a front-line role in making the changes happen. Once again, the same themes from earlier years played out with a small group of determined individuals affecting significant differences in outcome.

The interplay between the increasingly classified Space Control world and the emerging and free-spirited entrepreneurial space industry is an exciting story that involved my work at StratSpace. As the decade moved forward, StratSpace became involved in many highly classified space activities while at the same time keeping another foot in the venture-funded world. Ironically, the venture-funded space was bringing about a new space economy mirroring much of what was happening in the classified community at the time but operating faster and more innovatively. As with all such situations, people must eventually choose a side and put their efforts into one dedicated area. I decided, but not without a crisis of conscience fueled by my insight into the increasingly pervasive surveillance state created by some of my efforts and the ever-destructive forces of a foreign war.

One thing that continued after SpaceX was Cosmos One. Cosmos One was to be the world's first Solar Sail. My old friend Lou Friedman had made a deal with Ann Druyan, CEO of Cosmos Studios, who was also the widow of Carl Sagan. She liked flying a peaceful experiment, like a solar sail, out of a Soviet nuclear submarine on converted ICBMs that were once meant to destroy worlds. She called this "swords to plowshares" and was enthusiastic about the endeavor. We managed to find money from the Arts and Entertainment network A&E. They invested the money to finance this relatively modest program in exchange for doing a feature-length documentary film on the project. This seemed like a good deal when we started it.

Cosmos One solar sail experiment
Image Credit NPO Lavochkin

Before SpaceX, we had flown one Cosmos One experiment out of a Soviet submarine in Murmansk. It was a sub-orbital test to deploy one of the giant blades of the solar sail spacecraft. Our friends at Lavotckin in Moscow designed and built this experiment. We shipped it to Murmansk in late 1999 and installed it on an SSN18 with its warheads removed and our experiments put in their place. This flight was a success except for the terminal stage. The stage did not operate correctly and set the spacecraft on a trajectory where we did not know where it landed. The experiment had no telemetry and was to be recovered on land to recover the video and other data proving it had worked. As it turned out, somewhere in Kamchatka, our spacecraft landed and was lost forever. We still had the ultimate mission of flying the solar sail in space.

Lou Friedman conferring with Lavotchkin Director Konstanin Pichkadze
Image Credit Jim Cantrell

Once I left SpaceX, I could spend much more time and energy helping Cosmos One succeed. Lou Friedman and I made numerous trips to Moscow with a film crew in tow. It's easy to say that having a reality show made from your life is fun, but the reality is quite different. I did not enjoy it. I remember one time being very frustrated with the film crew. I turned to them and snarled, "Get that camera out of my ass." I was serious, but they thought that this was great cinematography. I found it simply embarrassing.

The work with our Russian colleagues on Cosmos One went relatively smoothly. It took several years to get the Solar Sail put together, but by the time 2004 rolled around, we could visit it in the cleanroom at Lavotchkin. It was a unique and unusual design. It had inflatable tubes which deployed a thin solar sail material. Each of these tubes was about 50 feet long. This made for a gigantic solar sail spacecraft. The blades themselves formed small vees much like a fan that you might find in a home. They were designed to rotate about their long axis so that depending on the sun's angle, we could spin the solar sail up or slow it spin down. This would give it stability in space. It was a surprisingly sophisticated spacecraft for just the $4 million we spent.

Cosmos One solar sail experiment in clean room ready for flight
Image Credit Jim Cantrell

By the spring of 2005, our Cosmos-1 spacecraft was ready to fly. We had the launch vehicle ready to fly out of a Delta Three nuclear submarine. We shipped the spacecraft to Murmansk and integrated it into the SSN18. The spacecraft replaced the nuclear warheads that had been previously installed. Interestingly, we could see in the corner of the warehouse where we were integrating the spacecraft the group of atomic warheads surrounded by soldiers and covered with a tarp. It was as if we had to ignore that existence. We tried not to look, but being close to nuclear warheads meant to destroy all of us was disconcerting.

The launch of Cosmos One took place in June 2005 from the Russian submarine Borisoglebsk. It began smoothly with the first stage ignition, and the launch vehicle sent along a typical trajectory. Friedman had placed himself on a boat in the Barents Sea very near where the launch took place. He photographed the launch vehicle exiting the water and ascending through the clouds. The first stage flight lasted about two minutes, after which the stages separated, and the second stage was ready for ignition. Things did not go so well once the second stage began to ignite. Based on post-flight reviews, the second stage exploded immediately upon ignition.

Cosmos One being loaded into Russian submarine for launch
Image Credit NPO Lavochkin

This was devastating for us. We had worked on Cosmos One for at least eight years and now had nothing to show. To make matters worse, the Russians were very secretive about what happened with the rocket. This is not surprising because the SSN18 was an operational launch ICBM and the Russians depended on it to provide strategic deterrence against a US nuclear attack. Naturally, the Russians were sensitive about such a failure. I consulted my sources in the United States Air Force Space Command and eventually got access to the classified data on the launch. It was clear that this vehicle had experienced a huge explosion during the second stage ignition, just as the Russians had claimed. That was about all I could proclaim publicly.

Lou was never able to recover emotionally from the Cosmos One failure. Several years later, the Japanese launched a solar sail that took away Lou's title of being the first solar sail in orbit. Clearly, the rocket provider, Makeev, had given us two bad rockets. Lou, for once in his career, had become disenchanted with the Russian Aerospace industry. I had a different experience than Lou before this and already had my share of disenchantment. We decided that the next time we tried some experiment like this, we would do it without the Russians.

However, like all of us in this business, we were still trying. In 2007, I came up with the idea of building a CubeSat and putting a smaller solar sail on a tiny satellite. This would be lower in cost and more straightforward. We were partly inspired by what NASA had done, something they called Nanosail D. This system was never designed to operate in orbit. Still, it was intended to demonstrate that a CubeSat could stow and deploy a small solar sail system. We worked with NASA to try and fly the remaining Nanosail D satellite but were unsuccessful in our discussions with them.

We began talking to his numerous donors to The Planetary Society, and Lou finally found a potential donor. We only knew him as "Richard," and he was willing to put up the money for the new Solar Sail. We called this program 'LightSail.' The overall system would be managed and designed by StratSpace, and various vendors would provide system elements. I contacted some friends in the Air Force research laboratory who had booms that could work for a Solar Sail, and they were interested in cooperating. I also reached back to my old friend Tomas Svitek. I first met Tom the summer I was at JPL working with Jim Burke. Tomas, as we knew him, had just arrived from Czechoslovakia. He and his family had escaped the communist regime by feigning a vacation to Yugoslavia. From there, they hiked over the mountains into Austria and spent several years in a refugee camp. Lou Friedman and Bruce Murray from Caltech sponsored Tomas to enter Caltech and pursue a degree in planetary geology. Tomas was less interested in geology, as it turns out, than he was in building spacecraft and other pieces of hardware. By 2001, he had started a small company called Stellar Exploration in San Luis Obispo, California. He was the perfect partner for us in designing and building LightSail.

Fathers of Lightsail from left to right Richard, Lou Friedman,
Tomas Svitek and Author, Photo Credit Lou Friedman

The year 2007 brought about some huge changes in my life. Due to long-simmering problems in my marriage, I faced a divorce. This was something I never wanted to do in my life as I was the product of divorced parents myself. However, it was clear to me that, given my circumstances, I had little other choice given her infidelity and refusal to stop. Further, I had to be a good example to my children and show them that we don't stay with people who are abusive or not suitable for us. It didn't matter the amount of suffering you had to endure. You must stand up for yourself and do what is right. This experience was very formative in my life. Not only did I have to start all over financially, but I had an opportunity to build my life anew. I have done many things that I was not proud of, and I now wanted to live a life in the future that I can be proud of and call my own.

As it turned out, I met my future wife in 2007 and didn't know it yet. When I was living in Key West and after my first wife had left me alone with two young girls, I had been out exploring the bars one evening with my friends. We went to one particular bar called the Bull and Whistle. The third floor of this bar was clothing optional. My friends decided they wanted to go up to the top of the bar and see what was happening. I had a slight interest in that sort of thing then.

I sat downstairs, ordered a beer, and watched the stage performance of a lady singing with a guitar. In my line of sight was the bar itself and a beautiful blonde woman who happened to have very long hair and was smoking a cigar. I found this very charming and unusual. I watched her for a while as she ignored her nearby boyfriend. It was an amusing and unforgettable scene. Very soon, my friends came back from the third floor, mentioning there was nothing to see up there. We went to the next bar, and I forgot about the incident.

I moved out of my Key West home in May 2007 and took my girls back to my Utah home. After returning to Utah, an old friend from Raytheon, Chris King, called and invited me to help write a substantial NASA proposal in Tucson, Arizona. As I made a briefing to this group, I couldn't help but notice that the woman sitting along the back wall was the same woman I had seen in Key West earlier smoking a cigar. I approached her during one of the breaks and told her I thought I knew her from somewhere. She felt that I was being gauche and trying to ask for a date during a business meeting. Nothing could be further from the truth. I was merely following a sense that I had seen this person before, and I had to put to some conclusion this sense that I had.

When I asked her if she had been to Key West recently, her eyes got huge, and she commented, "Why yes, I was!" I mentioned that I had seen her smoking a cigar at a bar, and her eyes got even bigger. We laughed about the experience and looked forward to working together that summer. We were simply colleagues for the rest of the summer, but eventually, by the time fall came around began to date and ultimately fell in love. Angela would eventually become my second wife by 2009. Arizona became a natural place to live since I had two young girls to take care of, and she also had two young children. I moved my businesses from Utah to Arizona and remain there still today.

We continued working on LightSail through about 2012, passed the Critical Design Review, and built both spacecraft. The two spacecraft had undergone a testing campaign and were awaiting a launch. Lou Friedman had decided to retire from The Planetary Society by this time. He originally asked me to take his place, and while I was flattered, I was more interested in continuing to build hardware. Ultimately, The Planetary Society hired Bill Nye, a well-known television star named "Bill Nye The Science Guy." Even though he was nowhere near a scientist by education or practice, this name from his TV persona stuck with him, and he was seen as a legitimate scientist. I got along fine with Bill initially. However, he is an intensely political person, and I am not. There was tremendous friction between us regarding politics, as I did not see the world through his eyes. I occasionally made fun of him that he was trained as an engineer yet sold himself as a scientist, which didn't help our relationship. He tended to opine on things he knew little about, which bothered me. These opinions also included LightSail and the business matters surrounding it.

One of the more significant questions that arose late in LightSail's development involved the intellectual property of the spacecraft we paid to develop. We had approached Cal Poly San Luis Obispo and their CubeSat lab to design the spacecraft computers and radio system. The Planetary Society paid for this new development which turned out to be very advanced and capable. Technically, The Planetary Society owned the intellectual property on this spacecraft design. However, several individuals from Cal Poly started a new company called Tyvak with this technology. They were going to commercialize it and sell spacecraft based on it. A considerable debate ensued within The Planetary Society about what to do with intellectual property infringement. I argued that enforcement of the IP ownership and a licensing agreement was a perfect opportunity to create an endowment for The Planetary Society, which occasionally struggled and didn't have two nickels to rub against each other. On the other hand, Bill Nye argued that it was simpler just to let the intellectual property go to Tyvak with no compensation and perhaps find other IP elements in the LightSail design to license out. He wanted to shrink from a fight and was willing to give away the store.

My mouth often gets me into deep trouble in life. I tend to have no filter between my brain and my mouth. Such was the case when discussing this intellectual property matter with the entire Planetary Society board. Bill and I argued vociferously over email about the right approach to LightSail IP. Finally, in frustration, I wrote to the entire board and Bill mainly, "Bill, you may be the Science Guy, but you're also a Business Moron". I did not survive these remarks. It took about a week before Bill informed me that I was fired as program manager from LightSail.

Furthermore, he fired Tomas and Stellar Exploration several months later. Somehow, Bill managed to find several million dollars more and hired another group to go through the spacecraft design and find fault with everything we had done. They spent several million dollars validating what we had done, but now Bill's mark was on LightSail, much like a dog urinating on a fire hydrant. I watched as he gradually took credit for the entire program, and this behavior made most of the original team that built it sick to our stomachs. I eventually learned to keep my mouth shut and was happy simply knowing that LightSail was my idea and built by the small team of people that made it real. The rest, like Bill Nye, were merely hanging onto what we had already done and taking public credit for it.

Lightsail flight experiment
Image Credit The Planetary Society

LightSail eventually launched aboard the first Falcon Heavy from SpaceX. This was 2018. It was a spectacular launch, and it was put into orbit. The spacecraft worked spectacularly, with the Solar sail blades completely deployed and the design operating as we had intended. I was very proud of what we had done and started to see how many of the things I had started to come together to make this one mission happen. I was very proud and remain very proud of LightSail as one of my career's more significant achievements.

In the time between Cosmos One and LightSail, I began to work with many classified programs which involved space warfare. Most of what I did I cannot speak about here. However, it was a very active area of the space industry with lots of innovation, and many bright people were attracted to solve the problems there. During this part of my career, I led many inventions, some patented and then classified. It was a time when the space threats coming from the Soviet Union and China were beginning to evolve, and we needed to find a way to protect our space assets.

I believe deeply in this mission, and it is a significant area I seek to support as I move along to new endeavors. The worst thing I could imagine happening to this burgeoning economy we were creating would be warfare in space. This would create debris floating around in orbit that would destroy many satellites. Many of my efforts to prevent a war in space focused on items that would deter space warfare. This includes rapidly deployed launch, rapidly configured satellites, and an ability to launch such satellites within days rather than months after the need arises. This capability will deter an enemy from wanting to destroy our space assets as we could quickly replace them and have minimal effect on the outcome.

By 2005, we were aware of the emerging capabilities in China relative to offensive space warfare. We watched closely in 2006 as the Chinese began to get very close to destroying one of their satellites. On January 11, 2007, the Chinese finally succeeded in blowing up one of their weather satellites in orbit and creating a gigantic debris cloud. At the time, I was briefed on this test and was aware that we were monitoring it. We had been monitoring many other Chinese tests and knew they had been missed many times. This time was a "success". A few days after the test, I was contacted by a reporter at Space News who had somehow gotten information that this test took place. I was forced to deny any knowledge of it due to the classified nature of the data. However, it quickly became apparent that the White House had leaked this information to expose the Chinese for what they had done. Within a few days, the news was everywhere, and the Chinese essentially had blood on their hands.

What is interesting for me as I look back on this experience, working in space warfare, is that many of the individuals leading the creativity of the efforts were later to become extensive parts of the new space industry. SpaceX's success in launching Falcon One and their later success in creating Falcon Nine led to space being safe for investment. It's fascinating that once the Falcon One was successful, SpaceX discontinued its production to make room for a multi-billion-dollar contract from NASA for the Commercial Orbit Transportation System. This was, in essence, a replacement for the space shuttle, which had just suffered a catastrophic loss during reentry. This loss killed the entire crew and led to the termination of the space shuttle program. I think it was the right decision in hindsight, but few of us saw this day coming. Even fewer of us saw that SpaceX would inherit the human transportation mission from the space shuttle.

The COTS program allotted roughly $1 billion in public capital to SpaceX to develop a large vehicle and a crew capsule. The large vehicle was called the Falcon 9, and the crew vehicle was called the Dragon. This private deployment of public capital was revolutionary. At the same time, NASA was trying to accomplish the same goals with a parallel program they had already spent $50 billion on by the time SpaceX succeeded with the first Falcon Nine and Dragon. This 50 to 1 ratio of capital efficiency demonstrated to all who would open their eyes to see that the future of space agencies and military agencies building hardware was over. Given our ever-increasing budgetary crises, these agencies could ill afford this kind of inefficiency and would eventually have to turn to the private sector to accomplish their core mission.

As the war continued in Afghanistan and Iraq, I began to think more and more about what was happening there. My son Colin had many friends I had known since they were little boys. They were young men joining the Marines, Air Force, and Army and heading off into the theater to fight these wars. Several of them would return home after having proudly served our country but with wounds to their bodies and souls. Some of them were missing body parts.

Others were so emotionally wounded that they took their own lives. I began to feel personally responsible for these young men, having played an enthusiastic part in the military-industrial complex for so many years and being an early cheerleader for the Afghanistan war.

Although I was only a tiny part of a massive war effort, I began questioning my entire career over the past ten years. If I honestly did not believe in this war and the government supporting all of us, then why was I in the business of war? As I pondered this question more, I came to the inevitable conclusion that to live what I considered an honorable life, I would have to walk away from the career I have known most of my life. At this point, I decided to walk away from aerospace in its entirety.

I didn't know what to do, but I thought I would quickly figure it out. I knew I had a thriving business called Vintage Exotics, where we built and restored vintage race cars. I could expand this business more. I could also return to my love of racing. And perhaps there would be other opportunities along the way that I could find. So, I returned to racing.

I had hired into Vintage somebody who became very much like my brother. Barry Ellis had built a local racetrack, Inde Motorsports Ranch, for a wealthy owner. I met him at the racetrack after becoming a member. When Barry disagreed with the track owners and left, he came to work for me at Vintage. That's when the magic happened. We made money buying and selling race cars, restoring old race cars, and repairing such vehicles for wealthy clients. We worked on everything from vintage Can-Am to Texas mile cars seeking to exceed 300 miles an hour in the running mile. It was a nice change from aerospace. We also started a race team. Barry and I won every race that we could afford to win. I began to have dreams of building a LeMans team and finding the sponsorship for it. The dream eluded me, however.

Barry and Jim working on CanAm
Image Credit Jim Cantrell

In 2010, I was approached by an old client from the aerospace world who had started a biometric fingerprint software company. Their former CEO failed to get sales moving and the products into production. They asked me to come in and do a turnaround on the company. This company, called IDAIR, had developed a novel touchless fingerprint imaging technology. I saw much promise and this technology, so I agreed to become a CEO. However, on the first day on the job, one of our clients hired our best programmer away from us. The person that hired our programmer was also our biggest customer. They had violated the terms of their contract with us. I got court action against them to prevent hiring this young man. However, their work together continued behind the scenes. They also sued IDAIR for patent infringement on patents that we owned. They claimed that they owned the patents by virtue of the contract work they had given us. We spent about a year in litigation with this customer but eventually prevailed. We prevailed because we discovered that they had been hiding evidence of their wrongdoing in collusion with their attorneys. We won the $17 million settlement in favor of IDAIR, and I decided that this year of fighting was enough for me. I had put their company back on solid footing, put a solid development team developing the product in place, and decided it was time to move on.

After my time at IDAIR, I began my professional auto-racing career. I started racing cars again in 2007, and the efforts were becoming more and more intense and competitive. Still, it was a part-time endeavor, and I had to share my time with more intellectual work. I had numerous sources of passive income and decided to pursue racing full-time. My goal was to eventually form a LeMans team and be one of its drivers. This was a goal well beyond my reach, and I knew it. However, one thing that my experience in the space business taught me was that I could never dream TOO BIG!

Barry and I built several cars to compete in various racing series. We started with some Corvettes for shorter road races and graduated to fire-breathing CanAm cars. We had numerous clients who would also have us race their cars for them as they enjoyed owning the vehicles but were often well past the point of running them. Our CanAm cars, one a Shelby and another a Lola, were successful in every race we entered them in. This was the first time that I exceeded 200 MPH while on land. Speeding around the Daytona Oval at over 200 MPH required the utmost concentration and skill. The track's curve gives the driver the impression that the road curves upwards ahead of you. As you exit off the banks and onto the straights, the car naturally skirts out towards the wall. As the vehicle nears the wall, you can do little to stop the centrifugal forces at 200 MPH from forcing this. The aerodynamic effects of being near the wall create the most unexpected result. The car 'bounces back' from the wall of air that you have created between the car and the wall. This 'cushion' gently pushes you back onto the track. The first time this happened to me, it scared the hell out of me.

Flying in the CanAm
Image Credit Barry Ellis

As we continued to race, I used the great advice of my driving coach Don Kutschall. Don is an intensely likable man with an impressive resume in racing. Don always carried an affable smile yet had a passion about him that was impossible to ignore. His shaved head and his penetrating deep blue eyes always got my attention when he was lecturing me about my driving techniques. He was also known as 'Dancing Don' on the track, as he used to be a professional dancer. Indeed, there are many connections between the two endeavors. At times, I have felt like the machine is an extension of my physical being, and driving on the track surrounded by tens of cars at high speeds gives me the sense that you are dancing with the others. This is where I earned my racing nickname, 'Dances with Machines, ' due to my skill at maneuvering through a crowd of cars and seemingly not touching one. At times I am close enough to another car to reach out of my window and turn down their rearview mirror but somehow, as if the hand of almighty God is at work, we never touch.

As we continued our racing efforts, we began to attract some serious driving talent to our ranks. Elliot Forbes Robinson and I began co-driving in several races. He became a mentor for me and often would take command of the radio while I was driving. I remember one race at Daytona where he took the crew chief position from Barry. I was leading the race, but I was not living up to the car's potential. EFR, as Elliot was called, had driven the car the day before to a lap speed of 1 minute 42 seconds. I was "lollygagging" around the track at 1 minute 55 seconds. EFR got on to the radio and yelled at me. "What the hell are you doing, Cantrell? Going out for a Sunday drive? Dammit, man! Put your foot in it. The car can handle more speed through the oval". I responded in the affirmative and poured on the gas. The car did handle it, and I was hurling through the banks at over 200 MPH despite my better judgment and instinct. As I exited the oval, the car skirted toward the wall, and I encountered the familiar aerodynamic wall. EFR returned on the radio, encouraging me, "That's what I am talking about, Son!".

He continued to tell me where I could find the grip on the infield corners, and I followed his advice. The next time I came around the start-finish line, I laid down a 1-minute 47-second lap. I received small praise from Elliot on my new time, but he expected even more from me. I followed EFR's advice on the next lap and kept my foot on the floor on the oval. The speed was intoxicating, yet I felt more and more in control. I ran the inside of the corners in the infield, going against all intuition and training, and came out on the last stretch of the oval and into the tri-oval at well over 200 MPH. EFR called out my final lap time of 1 minute 45 seconds. I received high praise for this being 3 seconds behind EFR, but I knew something had awakened in me during this race. I was now indeed among the ranks of the greatest. I didn't have the talent of many of them, but I could stand tall and count myself among the legends of racing.

Author and Elliot Forbes Robinson in Winner's Circle at Daytona
Image Credit Bob Johnston

Barry and I ran the CanAm in several historic races in 2011 around the country and began winning every race we entered. There is a magic between the crew chief, the mechanics, and the drivers. It's a bond of trust that is not replicated anywhere else in the world. In my case, Barry was both the Crew Chief and Chief Mechanic. I relied on him to prepare the cars for battle and my life depended on him doing this conscientiously and without error. This is a sacred bond between men much like those who have gone to war. My Seal Team friends tell me of the bond of brotherhood that forms between them in battle and how nothing breaks this bond. I began to understand this through my relationship with Barry. Because of this bond between us, I was thus free to concentrate my entire mind and body on being one with the car and winning. Barry and I were a magic team, and we both knew it.

Our ambitions in racing continued to grow. We had gotten a taste of endurance racing, and I decided that we were going to contest the epic 25 Hours of Thunder Hill. This was billed as the longest race in the world and was held every December north of Sacramento. It was almost always cold, sometimes clear weather, sometimes rainy, sometimes foggy, and occasionally snowy. We decided to build a car specifically to compete in this race. We chose a late-model Porsche of my design. I began with a solid streetcar our teammate Doug Nelson provided, and I stripped it to the bare chassis. I built the roll cage to the racing body's specifications and rebuilt the car mechanically from the ground up. Building an endurance car is all about reliability and making it last. You must keep it light and fast enough to be competitive. It was a natural strategic build and brought together a lifetime of knowledge and skills from being a line mechanic, building other race cars, and even building rockets and satellites.

Our entry into Thunder Hill was a success and became one of my life's crowning achievements. It's hard for those who have not experienced such an event to understand how much personal pride results from even finishing. We finished 6th in class and 42 overall out of 91 cars. At this point, it's important to note that our competition included ranked amateur teams like ours and off-season LeMans and NASCAR teams out for a 'fun weekend.' Barry served as our crew chief and didn't sleep for about 36 hours. He was both wired and dedicated to our team's win. I was one of five drivers, but one had to drop out at the last minute due to medical issues. We each took the car for two-hour stints. We were the slower cars out there, but our ability to go long on a single fueling was part of our strategy to place well. The strategy worked. The only real problem was that we had to drive defensively with the faster classes coming up on us at closing speeds approaching 75 MPH. One wrong move, and we wiped out our car and theirs.

We had done extensive practice leading up to this effort. Fortunately, I had access to a private racetrack at Inde Motorsports Ranch, where I was a member. Located an hour from Tucson, this track offered the perfect place to shake down the car. I was friends with the owner Graham Dorland and always had more freedom to use the track than most. We started by getting our pit crews out for training while we ran the car for a simulated 24 hours. We had decided to do this instead of waiting until we arrived at the races to see 'how it goes.' I learned from my years of building satellites that you 'test as you fly.' Here we were 'testing as we would race.' This test session gave us much confidence in the car, and we learned many things.

What remained for the entire team was a need for more experience in racing at night. I had devised some unique headlights and auxiliary lighting for the car (which later became products for my company Vintage Exotics), but we needed to test them. Rather than seek permission, we stayed at the track late and hid in the garage until everyone went home. Once it was dark, we set out on the track and began driving in the dark. My first impression was that it was a completely different place at night. The shadows transformed the surface into a very unfamiliar place. However, once we got used to the lighting conditions, we all grew more comfortable running fast at night.

Porsche 944 endurance car
Image Credit Jim Cantrell

One wrinkle, however, was a neighbor of the track who became very annoyed at us running with very bright lights shining like a nuclear power plant in the desert all over the hills and scrub brush. Since the track was all locked up, he couldn't get in to stop us. He also tried calling the police, who were not the least interested in interrupting whatever legal activities someone on private property was doing. He did manage to deploy a bright spotlight and tried to blind us at certain corners. The rest of the race team walked up and confronted him, finally sending him back to bed at about 0300.

The desert animals were the only other unexpected result of our night driving training. By now, we were in full summer, with most animals being nocturnal and active. The bright lights attracted a lot of wildlife to see what was happening. We could spot coyotes on the side of the track, Javalina (miniature wild pigs), rabbits, and snakes. The rabbits were the most curious of the creatures. During one of my stints, a rabbit decided to come onto the track at the end of the fast long straightaway. He placed himself at my turn-in point at the edge of the track, which meant I had to pinch the corner by moving more to the inside of the turn to avoid hitting him. Every lap, he got further out into the track. I finally decided not to move and expected to scare him off the track altogether. He didn't move either and our collision created quite a mess all over our white car. It was somewhat of a horror show as I showed up at the garage with a car covered in blood. A quick wash and a few minor repairs had the car back on the track and we drove until the sun came up. It was an eventful and worthwhile night, but we paid hell from the track owner for taking such a liberty. It was still worth it.

Our race at the 25 Hours went well. We had a strategy of the tortoise and the hare. Let the other cars be the hare, and we would be slow and steady like the tortoise. Slow, mind you, were still 85 mph average lap speeds with top speeds above 140 mph. We ran clean stints and stayed out of trouble for most of the race. I was the only one to get hit. I insisted on taking the mid-afternoon as my first stint, the next one at one in the morning, and then sunrise. Such is the nature of my biorhythms. It was at about 0200 in the morning as I was cresting a hill that a car came out of nowhere and hit my front left fender. Instead of pushing the car out of the way, the other vehicle, an Acura, drove right up the side of my car and went on two wheels. I could see the other car's underside with my excessively bright lights. He managed to control the vehicle and continue as if nothing had happened. I ended up sideways coming down the hill and somehow managed to hold the car at 100 MPH in the dark. When I could see the track again, I slowed quite a bit, but the car still seemed intact. I radioed in that I had been hit and was coming in for an inspection. They looked at the car, and nothing in the way of damage could be seen besides some wicked tire marks: those damned German engineers and their Porsches.

A completely new place revealed itself as the sun rose over the track. It resembled more a battlefield than a track. There were burned-out remains of cars, bent-up hoods, and broken exhaust systems off to the side of the track. It was as impressive as it was amusing. I finished my final stint at 0700 and turned the car over to the rest of the team to finish. I slept the sleep of the just and woke up a few hours before the finish. The completion of the race was glorious, and we ended up finishing very well. After packing the car and pits, we all returned to Arizona. I slept in the truck's cab for the first four hours. Barry slept in the back bed of the truck under the camper shell and managed to sleep the entire journey to Arizona. We dropped him off at his home just as the sun was rising. It was a glorious achievement, and we all knew it. We didn't have to talk about it. The evidence was in the doing. I could not have been prouder of my team and myself.

By 2012, SpaceX was very successful. Many people were working to replicate that success with private capital in the space industry. It was still very much a frontier economy. I started to see efforts arising out of people coming out of the classified military world to create companies that would do commercially what had previously been done in the classified setting. One such company was Skybox Imaging. They approached me very early after leaving the NRO and told me about their plans to produce a constellation of spy satellites for $1 million per satellite. At the time, satellite costs were hovering around $10 million on average for a small satellite. A suitable imaging satellite capable of producing 1 m or less ground resolution should easily cost $50-$100 million each. Skybox had lofty goals, and they wanted to know if I thought it was possible. I studied and decided that while $1 million may be too ambitious, we certainly could build satellites for $5 million each.

Over the next several years, I helped Skybox raise several private financing rounds. I also help them assemble a team of reviewers and engineers to design and build the satellites. They were very successful in making this happen. It was a great team and a great idea. They flew their first few satellites on a Russian rocket harkening back to my early days with Kosmotras. I had returned to the Bear for Skybox to negotiate these contracts for our flights from the same people who had turned Elon down. We encountered a completely different reception from the Kosmotras people after I returned the second time. This time they took me and my negotiating companion seriously. We secured a ride on the decommissioned ICBM to launch out of the Dombarovsky Air Base in Russia. By 2014, Skybox had raised over $70M and had its first satellite in orbit.

Skybox ended up attracting some of the best talents from the industry. We managed to recruit the former Director of the Goddard space flight Center at NASA, some people who had initially built the Digital Globe imaging systems, and many others in the industry whom I consider the best entrepreneurs and engineers. It was clear from this experience that the private money poured into it would attract the best and brightest.

This caught my attention and gave me many thoughts about the future. If I had any doubts before this, they were all erased on the day Google offered to buy Skybox for $500 million in cash. The sale went through, and the company was transformed by its ownership by Google. Only some of those effects were good. I remained attached to Skybox for several months after the acquisition. However, the environment changed completely. Google was, in essence, coddling the Skybox employees and, in my opinion, "overfeeding them." Google would have every breakfast, lunch, and dinner catered to the crew to encourage them to remain on the job. This created a culture where everyone wondered what the next meal would be and spent an awful lot of energy wondering and preparing for that meal. Ultimately, Google ended up selling the Skybox franchise to Planet Labs. Planet Labs was also a start-up formed with venture-capital money and had several hundred small CubeSats already in orbit. The addition of the Skysats from Skybox allowed them to produce higher-resolution imagery.

Skysat Satellites ready to launch
Image Credit Skybox Imaging

Despite my highly successful racing career, I was convinced to join the ranks of a company called Moon Express. Moon Express was venture-funded and aimed to send a privately funded lander to the moon. This company had already attracted great talent, including Alan Stern, the PI for the Pluto flyby mission and former NASA associate administrator for space science. Other great engineering talents were attracted to the company as well. I came on as Chief Technical Officer and took a role in the overall architectural design of the lander. We raised about $40 million for this effort while I was there.

Our initial efforts focused on a lander design based on a Toroidal propellant tank, which was technically challenging to implement. After about two years in this new job, management problems within the company caused an engineering revolt during my time there. Many original engineers left the job, and we were stuck rebuilding the team from zero. As a result, I led an effort to redesign the lander to be compatible with a new class of emerging small launch vehicles. A significant problem with the earlier designs was finding a launch that we could afford. Most launch opportunities could take us to the Moon by piggybacking on a satellite launched in a Geostationary orbit. The cost of such a launch for us was around 30 million dollars.

Rocket Lab, a New Zealand-based company, was offering a smaller vehicle for $6 million. Several years earlier, I had done some work on behalf of Bessemer Venture Partners to evaluate Rocket Lab and concluded it was a very viable design. I did some initial calculations and discovered that a small lander could be launched on the Rocket Lab Electron. The so-called micro lander I had conceptualized would be launched as a dedicated lander on the Electron. I got approval from Bob Richards, the Moon Express CEO, to go ahead with this design, and my longtime colleague Bud Fraze and I began to work on it.

Bud and I formed a magical team like Barry and I had before in racing. Bud had a long experience designing everything from nuclear warheads to Fusion power plants to the initial Digital Globe satellites. I was good at systems engineering and could quickly size and specify the lander and iterate with Bud on the design in real-time. Within about two months, we had what we called our MX micro lander design nearly complete.

Bud and I continued working on this and held several reviews with NASA and outside reviewers. We had a contract with NASA to provide a commercial space mission to the moon and Bob was busy raising the money to front-load that mission. We signed launch agreements with Rocket Lab and were ready to begin constructing a prototype of the lander. However, one of the major problems was that at this point, we had spent so much money on the prior design, which was a failure, that few investors wanted to put more money into the company despite being on a better technical footing.

Ultimately, Bob could not raise enough money to build the MX micro lander, and Bud and I decided to part ways with Moon Express. Bud and I remain colleagues; he has worked with me on all my ventures since then. I also met Mike D'Angelo during my time at Moon Express, and he later became my business partner in Phantom Space Corporation. Like many other startups, they don't work out, but you learn from them and find people who become business soulmates. Mike and Bud were the two I took forward in my future endeavors.

By 2015, other startups approached me regularly to help them get started. One of my favorites was ICEYE in Finland, which decided, like Skybox, to build military-quality imaging satellites for single-digit millions of dollars instead of the single-digit billions of dollars the government paid to Lockheed Martin and Boeing. These radar imaging satellites that ICEYE was producing would cost about $5 million and could produce imagery equal to what the spy agencies regularly used. The CEO of ICEYE, Rafal Mordiewski, is a brilliant young Polish student who ended up in Helsinki, Finland. He had met other students there and decided to start this company.

Iceye satellite
Image Credit Iceye OY

Some of my early advice to Rafal was that building such a satellite would create an enemy within the US intelligence community. I had spent my time there and seen many satellite programs for radar imagery that would come to life and mysteriously die. It seemed as if there was a hidden hand in the background of all the intelligence agencies that would seek to destroy government and private programs that would challenge the US government's secret dominance in radar imaging systems. I would find out later why this was so, but I warned my colleagues at ICEYE about this as a potential outcome. However, since this company was based in Finland and Finland was considered a neutral country then, this was probably the most likely place for such a company to exist without interference from the US intelligence community. I turned out to be correct on this, and today Iceye is a very successful company with seven satellites in orbit.

In early April 2015, I received an exciting phone call from someone who identified himself as Reuben Sorenson. He claimed that he worked for the Defense Department Joint Chiefs of Staff. He claimed that he was a science advisor to General DeSilva on the DoD Joint Chiefs of Staff. Reuben was calling me because he had read an op-ed piece, I had written for Space News calling the acquisition systems of the defense department a "Soviet economic system." Reuben claimed to resonate with what I had written and was looking for help getting some inroads into the commercial space industry, where he had specific needs.

Of course, I am googling his name while talking to him and finding nothing about Reuben Sorenson. I told him on our phone call that he "didn't exist." He mentioned to me that he was a Seal Team member before this position. He was part of a group called DEVGRU, more commonly known as Seal Team Six. This entire story seemed like bullcrap, and I called him on it. Frankly, I suspected he was an FBI agent trying to set me up for something I hadn't done yet, and I was very reluctant to speak to him. I told him I would be happy to help him, but I would have to verify his identity. My days in the Soviet Union and Russia taught me a lot about personal security, and verifying people's identity before you talk to them was basic. The world is full of smoke and mirrors, and this seemed to be yet another case of it.

Reuben seemed willing to identify himself openly and invited me to come and meet him in the Pentagon and have an introduction to General DeSilva. I agreed that this was a good idea, and once I had verified his identity, I could help him. Reuben described his problem as involving North Korea, where the defense department was unaware of pending missile and nuclear weapons tests. Reuben told some of his work in Iraq, where he went into Falluja to help stop some IED attacks on US soldiers. He had set up what he termed a "neighborhood watch program" using UAVs and commercial imagery that he had placed on these aircraft. Admiral McRaven gave him this challenge, and Reuben was sent to Iraq by him to solve this problem.

Once he arrived, he discovered that the military UAV imagery was classified upon receipt and sent directly to the intelligence agencies in Washington DC. It was later disseminated to ground troops on a need-to-know basis. Reuben hypothesized that by setting up a more responsive imagery collection system, he could detect the attacks and trace them back to their source more readily. His approach was to tape on GoPro cameras literally to the fuselage of the UAVs, and once they landed for refueling, remove their SD cards and process and imagery. I imagined the response of the Air Force mechanics when he began duct-taping the cameras on the fuselage and the kind of stir this must have made among the rank-and-file commanders. This approach, however, led to great success.

Reuben told me of an IED attack where they could run backward in time with the archive footage and see exactly who had placed it and where that vehicle had come from. It led to a small building on the outskirts of Falluja, where the bomb makers resided. Air raids by F-18s eradicated this problem. Except for the F-18s, Reuben wanted to do the same thing in North Korea, but he needed to use commercial satellites to do this. The problem was, as he explained, that the US intelligence community did not take adequate images of North Korea, and the ones they did take were done at such a predictable time that the North Koreans would hide their missile and nuclear activities from the imagers. On the other hand, commercial companies were not interested in taking images of North Korea because there was no real market for it. Our challenge with Reuben's new program would be encouraging commercial image providers to provide imagery of this reclusive and otherwise uninteresting country.

I consulted an old friend, General Mike Carey, to verify Reuben's identity. Mike made three stars as an Air Force general and was at one time in charge of all American nuclear weapons throughout the world. He's now retired but maintains excellent contacts within the Pentagon. I started with him to verify Reuben's identity. As luck would have it, General DeSilva and Mike had graduated from the Air Force Academy together. It was simple for Mike to call General DeSilva and ask about Reuben Sorenson. It turned out to be very comical that General DeSilva made it to Reuben's office before Reuben could get the invitation for me to visit. I received an incredulous call from Reuben about an hour after our first call, and his words to me were, "Dude, you're checking up on me already?"

After these series of calls, it was clear that Reuben was the real deal, and he was everything he said he was. I had doubted this man, but here I was, standing in the presence of a true American hero. At the time, I didn't know the full extent of Reuben's accomplishments within the Seal Team environment, but to say that Reuben is a national hero is to put it mildly. I volunteered my time pro bono and asked him how I could help.

Our first task was identifying the image providers relevant to Reuben's project, which he called Data Hub. As it turned out, my old friends from Skybox Imaging and Planet Labs had the most pertinent satellite constellation, and their satellites were in orbit such that you cannot readily predict the overflight times. It was perfect for what we needed from a technical point of view. The problem with Planet was its CEO, Will Marshall. Will is the very definition of the term "Peacenik." Will would not allow his constellations to be used for what he considered nefarious military purposes. Anything related to military or intelligence end uses qualified as "nefarious" in Will's world. It didn't matter to him if these applications were intended to keep the peace. It was more of a blanket attitude towards this entire end-use.

We identified a company that could take this imagery, process it, and turn it into intelligence products that the US Department of Defense could use. Orbital Insight was founded by a colleague from Moon Express named Jimmy Crawford. He had many of the algorithms developed for autonomous tracking and possessed the capability to deliver more. However, he needed the raw imagery to create the data of products that the intelligence agencies required. Reuben and I set about creating a false story with Planet Lab regarding the economic future of North Korea. We encouraged them to gather data regularly on North Korea and sell it to Orbital Insight, which had financial customers who were very interested in this data. This worked very well, as it turned out.

Right after the New Year in 2016, I sat at the computer in the evening to look at my email. One of the things I monitor is earthquakes around the world, and the US geological survey sends out emails for large earthquakes right after they detect them. One of my emails stated that there had been a 4.5 Richter earthquake in

North Korea that day. I forwarded this to Reuben and joked about converting the Richter scale into megatons. This was not a very funny email as he delivered it to General DeSilva and others on the joint Chiefs of Staff who needed to be made aware that North Korea was preparing a nuclear test, let alone had conducted one. Again, before we had a chance to have Data Hub operational, the North Koreans had done something that surprised us all. As it turned out, I was the first among anyone in the Pentagon to know about that North Korea nuclear test. Such is how the US intelligence agencies have decayed in their ability to use modern technology and be at the forefront of knowledge.

Reuben's Data Hub program began to have great success in 2016. They identified certain behaviors in North Korean society that could be observed from space that correlated with both missile and nuclear testing. For a mere $10 million, the US intelligence agencies and the Department of Defense now had a program where they could predict North Korean behavior. This was a brutal awakening for many in this world. It showed how wasteful much of the defense spending could be and how truly focused it remained on a historic enemy such as the Soviet Union. In this new world, we needed to be able to exploit information rapidly and the defense department was not doing that particularly well in the space business. Reuben received multiple awards for the Data Hub program, which continues to this day. I'm happy to have spent a small part of my time on this, and I'm very proud to call Reuben a friend and colleague to this day.

Article on small satellites and Datahub
Image Credit Reuben Sorenson

Following my work with Reuben, I begin to think about the problems of the emerging commercial space industry and how best to serve it. Through my experiences with both Skybox and Moon Express, it became very apparent to me that one of the major forces holding back the progress in this emerging economy was that of launch. In all the start-ups I had been involved with, the launch was always the Achilles heel of the business. There's not enough launch capacity for the small satellites to satisfy the demand. Furthermore, the available launch opportunities flew at an inconvenient time for most satellite operators and went to orbits where many of the satellites didn't find optimum operation.

I began thinking back to my original days at SpaceX. Chris Thompson and I both saw a future where the Falcon One could be mass-manufactured, and through mass-manufacturing, the cost of individual launches would fall due to the learning curve and supplier cost reductions. At the time, we did not see the emerging economy of a giant reusable rocket like the Falcon Nine. However, it was clear to us that mass manufacturing had a role in future launch architectures that would bring launch frequency up and costs down.

What began to emerge in my mind about 2015 was that SpaceX had begun to perfect the large vehicle using a reusable first stage, but it was only part of the launch solution. Reusability reduced the amount of time required to recycle a vehicle for launch. The result was that SpaceX could launch up to 60 of these large vehicles per year by amortizing standing army costs across a larger number of launches. There were additional cost savings by not throwing away the first stage, but the stage refurbishment costs offset these savings. In contrast, small vehicles could be mass-manufactured, and the individual cost of each vehicle would be successively made lower and lower through the effects of the 'learning curve'. Henry Ford and the Model T proved this and have become the standard for most industrialized products today. Thus, if the mass-manufactured small launch economic model proves viable, two competing economic models for launch will form and define the next 20 years of space transportation.

A result of my involvement in working with bankers to raise several billion dollars to resurrect the satellite phone company Iridium and later venture capitalists in other New Space companies, such as York Space Systems and Iceye, was the recognition that launch was a critical factor in the future viability of the space economy. Despite all that SpaceX had done for the cost of launch, it was also clear that Elon is headed for Mars with his Falcon vehicles and that the New Space community needed a transportation system to build the new space economy.

At the end of my self-imposed exile from the space industry in 2015, I re-engaged my old colleagues who helped start SpaceX, and we formed Vector Launch to fill in the space launch gap left behind by our friend Elon. In a nod to Henry Ford and the Model T, Vector sought to employ mass manufacturing techniques from the auto industry combined with modern materials and age-old rocket designs to create space launchers by the hundreds per year for 1/100th the cost of Elon's Falcons. Vector did not survive for several interesting reasons that had little to do with the fundamental viability of a small launch. I left Vector in August of 2019 due to disputes with my other co-founders, and one of them, John Garvey, assumed the position of Vector's CEO after I left. Unable to raise money, John led Vector into bankruptcy in December 2019 and began a liquidation process of its assets.

By late 2019, I had formed a new company called Phantom Space, which had an entirely new team, a more comprehensive business plan, a clean sheet rocket design, a refreshing approach to the supply chain, and, most importantly, incorporating the lessons I learned from Vector. As a result, I still find myself in the middle of what I am calling the rocket wars. I have dedicated the last five years to making this a reality, yet the story still needs to be written entirely. I will leave this chapter of the New Space story for a future body of work where I can tell the complete story. As of late 2022, Phantom is 12 months away from its first launch, has won several NASA launch contracts, and has raised over $100M. Its market touch is broader by providing both launch and the customer satellites that launch on our rocket known as the 'Daytona.'

Phantom Daytona Rocket at Vandenberg
Image Credit Phantom Space Corporation

CHAPTER 10

BURN THE SHIPS!

To our world today, space is much like the New World was to Europe 500 years ago. Standing on the shores of Spain at that time, few could have imagined what would become of the New World and the untold value that would be created there. Seeing the Spanish galleons leave to explore the New World went few with any idea of what they would return with. Even fewer could imagine this traffic multiplying many times over the next five centuries. In this faraway world, over the next five centuries, new cities rose, fortunes were made, and great new nations were born. Wealth beyond anyone's imagination was discovered in this New World. Even more, wealth was created by those brave enough to have ventured away from their predictable lives in Europe into this unknown new world.

Yet for all the wealth and the potential sitting directly in front of us, we humans cannot often see an enormous opportunity. A new frontier represents an opportunity for us as individuals, society, and species. Today, we struggle to imagine how we will derive ultimate value from conducting our affairs in space when doing the same thing from the comfort of Earth's surface often seems much more rational. This parochial small-minded view is reassuring that we are still humans and still are attached to our terre natale. It is, after all, the place that gave birth to us and has housed and protected us for millennia. But now we find ourselves, once again, at the very beginning of something huge, something that will be a sea change in human history. As we take our first baby steps off the Earth, we will undoubtedly have moments of anxiety and regret. But as with all of life, there is no choice but to grow and move. As Hernan Cortes said to his men when they landed in the New World in 1519: "Burn the ships!". There is no turning back.

Morgan Stanley predicts that by 2040 the world space economy will exceed 2.2 trillion dollars. To put this into perspective, this market size is magnitudes more prominent than the current worldwide auto industry and exceeds the net proceeds from the internet and computer industry combined. In our lifetimes, the space economy will likely become humanity's single largest economic sphere. Today's space economy is comprised mainly of imaging and communication satellites. As of 2022, the space economy is valued at $400B annually, resulting from only these relatively narrow space applications. The near-term space economy will continue to grow this essential infrastructure. Still, it will gradually include new and novel applications of data storage and computation in space and new ways to see and re-imagine our world. Self-driving and connected cars, remote networks of machines performing the toil and hard work once reserved for humans, and our homes will be linked together through space-borne communication systems bringing unimagined value and change to our lives. Much like what GPS has enabled in our modern world, the current generation of space applications will continue to innovate and create new and novel utilities from this vast medium surrounding us. Along with this innovation will arrive a whole new class of millionaires, billionaires, and our first trillionaires.

Like elevator technology making the Manhattan skyline possible, emerging space technology creates an entirely new physical layer from which we can create value for the rest of humanity. This elevated economic medium currently serves as an efficient channel to move bits around the Earth, take images of our planet, and otherwise turn photons into decision-quality information. These space-derived data shape our lives today with weather forecasts and hurricane warnings. Still, they will evolve into more value-added applications of these essential functions in the coming decades. We lack the imagination to conceive of how looking back at our Earth can be turned into value chains and human advancement.

The earliest uses of space imaging were aimed at weather observations and gathering helpful intelligence by competing nation-states. The earliest space images were returned to Earth as film canisters to be developed into full-sized paper images, much like early Kodak cameras. This technology remained state of the art for the highest resolution cases until the late 1970s when digital technology and our ability to transfer data improved. By the 1980s, nearly all imagery of the Earth taken from space was digital. This highly portable format opened a new set of possibilities to analyze and manipulate the imagery more quickly and with new computer algorithms. In short, digital imagery opened a new range of options, and the digital format enhanced its mobility to be sent worldwide instantaneously. Space probes sent out by NASA and the Soviet Union as early as the 1960s began sending fuzzy and faint digital images back from remote worlds like the Moon, Venus and Mars. By 1979, when the NASA Voyager spacecraft began flying by every planet in the outer solar system, the quality of imagery sent back for the world to see was stunning.

While the pace of focal plane and optical technology alongside computer processing power limited what could be done with the imagery in the 1970s, those technologies advanced, and by 1990, many of the premier digital focal planes and computers to process the images were available commercially. Still, by the early 1990s, high-resolution images from space remained in the government domain by decree. While it was technologically and fiscally possible for private companies to send high-resolution imaging systems into orbit, world governments restricted such activity through restrictive laws and regulatory schemes. In the 1980s, it was literally illegal for a private company in the US to build, launch and operate a commercial imaging satellite from orbit. Such was the level of secrecy and protection given the cold war state. The names of the agencies that built and flew them for the US government were classified. It was not long ago in my lifetime when I could not say or write National Reconnaissance Office or its acronym NRO openly. We euphemistically referred to the agency as 'Westfields' based on the local name of the area where the agency was located, hiding in plain sight. Likewise, the National

Security Agency's name and its acronym were highly classified and often referred to in jest as "No Such Agency".

Little by little, these barriers were broken down, and the first commercial Earth imaging satellites were launched by the end of the millennium. By design, these commercial imaging systems were limited in resolution, and their ability to be useful as intelligence-gathering devices were blunted, or so thought the nation states restricting this technology. The US government maintained what it termed 'shutter control' on commercial imaging satellites by placing regions of Earth and sky off limits to imaging and, in so doing, felt more secure in allowing the commercial imagery market to develop. In a perverse turn of fortunes, however, retail remote sensing data and imagery meant for commercial purposes were eventually found to have immense value for intelligence gathering. They began replacing government satellite imagery after the turn of the century. What used to require billions of dollars to the field by the US military was now replaced with commercial capabilities that were as good or better than what they replaced. The retail remote sensing industry that began in the 1990s roared forward with great optimism. Still, commercial demand for high-resolution imagery was sorely lacking by the latter part of the decade. There were novelty applications, of course, but the forecast uses of the imagery for precision agriculture and Earth mapping was well ahead of its time. Ironically, the US intelligence agencies found a great application of this imagery to supplement its other capabilities. Unique surveillance methods from space were often termed 'exquisite measurements' in bureaucratic parlance but referred to other than 1m class visible imagery. However, a protracted crisis in developing future visible imagery systems for this community created the perfect opportunity for the struggling commercial space imaging industry.

In a pattern that would be repeated by almost every other government agency, including NASA, the National Reconnaissance Office (NRO) had been developing a new generation of visible imaging systems to replace the aging KH-11. NRO selected Boeing in 1999 as the prime contractor for the Future Imagery Architecture (FIA) program. FIA aimed to replace the KH-11 satellites with a cost-effective constellation of smaller, more capable imaging satellites. FIA was awarded to the Boeing Company with a total budget of $10B. By 2005, an estimated US$10 billion had been spent on FIA, including Boeing's accumulated cost, overrun of US$5 billion, and estimated to have an accumulated cost of US$25 billion to complete.

In contrast, each Worldview satellite could be built and launched commercially for around $500M-$800M. FIA was canceled, and the NRO procured nearly 80% of the capacity of the US commercial imagery market. This was a rather significant shift from nation-state domination to the US government becoming more mission-oriented and less focused on developing what the commercial industry is more efficient in doing.

The current generation of space technology continues to evolve rapidly and in the coming decade will continue to impact our daily lives creating untold value for investors, employees, and ordinary people. Technologies currently under development will be employed in space systems and will change the face of our global economy. The near-term space economy will generate new ways for communications to reach the most untouchable parts of the world, bring knowledge and enlightenment to places where governments try to keep citizens' access to such information limited and allow us to watch and observe our beautiful blue planet in ways that we have never imagined. Much of what we touch daily in our mobile phones already relies on space assets, but the future will entirely rely on space assets and the industries that produce our space technology.

In many ways, we owe a debt of gratitude for this space technology revolution to Elon Musk and his pioneering efforts with SpaceX. Elon's effect on the space industry and the venture investments going into that industry are hard to understate. Many say that SpaceX's success made space safe for investing again. The latest wave of investments began ten years ago and since that time, 30 billion dollars in venture capital and significant investment funds have placed bets on this emerging economy. Something is stirring in this budding industry, and it is being fueled by a new generation of entrepreneurs and investors willing to take risks alongside these entrepreneurs. Commercial space is nothing new, as space investments have occurred for over 40 years. This latest wave, however, seems to have momentum that none of the prior investment periods experienced.

By 2004, global investment in all space sectors was estimated to be US$50.8 billion. As of 2010, 31% of all space launches were commercial. Today, over 80% of the satellites launched yearly are commercial and not government funded. Commercially funded space started arguably in the late 1960s and early 1970s with large communications satellites. These satellites derived from more extensive government-funded programs, and the price tag for such large satellites was generally in the $100M-$200M range. Global communications markets, along with television broadcasts and international phone service, were in their infancy. The first commercial use of satellites was the 1962 Telstar 1 satellite, the first privately sponsored space launch. AT&T and Bell Telephone Laboratories funded the satellite, and it was capable of relaying television signals across the Atlantic Ocean. This was the first satellite to transmit live television, telephone, fax, and other data.

The Hughes Aircraft Company developed the geosynchronous Syncom 3 satellite and leased it to the Department of Defense. Syncom 3 telecast the 1964 Olympic Games from Tokyo to the United States. This novel technology found increasing application, and commercial companies, such as Hughes, created what is now known as the commercial Comsat industry. Most of these satellites were placed in what is known as Geostationary orbits which have the characteristic that the orbital period is 24 hours. The net result is that the satellite 'hangs' over the same position on Earth. Communications satellites placed in a Geostationary orbit result in Earth-based antennas that do not have to rotate to track the satellites but remain permanently pointed at the position in the sky where the satellites are located. Slots in Geostationary orbit thus became much like real estate on Earth, with international agreements dividing the orbit positions to countries and entities.

The second wave of commercial space began in the early 1990s with the emergence of commercial imaging companies. Government regulatory reform made way for private companies to build commercial satellites and launch them. The 1992 Land Remote Sensing Policy Act (enacted in October 1992) permitted private companies to enter the satellite imaging business, forming many commercial remote sensing imaging companies: Earth Watch (now Worldview), Ikonos, and GeoEye. The most successful of these was Earthwatch which ended up consolidating with GeoEye. Many of my colleagues, including Bud Fraze, worked on this early spacecraft and continued bringing this pioneering spirit into the modern 'New Space' era.

Today we find that the space industry, once dominated by nation-states and nation-state-sized budgets, increasingly relies on a blossoming commercial space sector to supply the infrastructure that both commercial and nation-states can conduct their business. This trend is here to stay and is set to grow. Many fundamental changes are underway or have already happened that will ensure this new commercial reality will remain dominant for the foreseeable future. Structural changes have occurred that will only reinforce the role of commercially developed space systems for the future of the space economy.

The space economy comprises the range of activities and the resources deployed in managing and utilizing space. It includes public and private sector organizations that use space-related products and services ranging from research and development, the manufacturing and use of space infrastructure (ground stations, launch vehicles, and satellites) to space-enabled applications (navigation equipment, satellite phones, broadband or machine-to-machine communication services, space-based imaging systems, meteorological services, etc.). Many countries view the space market as strategically important from commercial services and national security perspectives, and the space market is expected to grow from U.S. $400B in 2020 to US$2.7T by 2045E (Source: BAML Space Report 2016). Private sector investment has also rapidly expanded, with $16B invested in space-related start-ups and established companies since 2000. $3.9B was invested in 2017 alone across launch and various elements of satellite hardware and services companies. (Source: Space Angels report).

A vital component of the space industry is the satellite segment. The market is moving towards smaller satellites. Well over 8,500 satellites are orbiting the Earth today, generally split between commercial, military, government, and civil (primarily academic) sectors. Applications are varied but can be broken into communications, observation, science, and technology development testbeds. U.S. citizens benefit significantly from satellites for everything from video/voice calls, navigation, inflight Wi-Fi, web maps, and driverless cars in the future to weather monitoring. To put the value of satellite services into perspective, GPS alone contributes up to U.S. $75B annually to the U.S. economy (Source: US PNT).

Small satellites are transforming akin to what the personal computer market experienced in the 1980s, which is driving radical changes in the space economy. As computing costs vastly decreased and computing capabilities increased, the computer user community broadly proliferated. The same effect is underway today in the satellite market, with ever-increasing numbers of small satellite manufacturers and satellites being launched. In 2020, a record 1282 satellites were launched into orbit, with over 85% being small satellites. Small satellites currently account for 20% of the up mass to orbit, which is expected to increase to over 50% by 2025.

The Falcon 9 has become the US launch workhorse
shown here ready to launch.
Image Credit SpaceX

The satellite market is at an inflection point, shifting away from large, expensive hardware to smaller, less costly small satellite systems that require a smaller, more cost-efficient launch capability.

This shift can be likened to what the personal computer market experienced in the '80 and '90s, driven by a decrease in cost and miniaturization, increased capability, and customer demand. "Nano" and "micro" satellites (weighing 1-50kg) are also increasingly popular but have more limited functionality and short mission life (1-2 years). Since the first "CubeSat" launched in 2002, the number of tiny satellites sent into orbit has increased rapidly. Continued advances in miniaturization and satellite integration technologies are expected to significantly increase the functionalities of nano and microsatellites and their possible field of application.

The speed of manufacturing satellites is increasing as well. Traditional communications satellites require as much as 36 months to build and integrate for launch. By early 2015, Planet Labs needed nine days to make two CubeSats. Satellites that used to take 2-3 years to design and develop can now be rapidly designed, built, and launched frequently to best respond to dynamic market needs and to take advantage of the latest technologies. Moreover, more satellites will be launched in the next six years than in the prior 50 years, and in 2021, over half the mass of satellites launched were so-called small satellites weighing less than 500 kg (1200 lb.).

Much of the work started by the 'lunatics' and 'space cowboys' thirty years earlier in small satellites has led to today's technology revolution. This includes cell phone technology being deployed in ever large numbers in ever less expensive satellites and is set to create a technology revolution equivalent to the internet of the 1990s. Many of the same individuals who showed up in earlier episodes in Russia, billionaire-funded space ventures, and military space programs came together in the past decade to renew the world space industry and propel humanity's future faster into space than anyone had imagined even ten years ago.

While past commercial space eras have often slowed and faded, the latest wave of commercial space investments has some important distinctions that will propel it forward with permanence, unlike earlier waves of investment. Looking back on the prominent LEO constellations and imaging constellations of prior eras, the most significant factor constraining its growth has historically been the large amount of capital required to make the business viable. There was little question that most technology was real and relatively low risk. The business case of commercial imagery and large constellations for data transfer, like Iridium, was far riskier. The result was, in the 1990s and 2000s, space companies trying to raise $1B, $2B, and 5B dollars to finance a business case in space. That money could only be raised for a very low-risk business case.

Today, due to other investments made into the space industry, the satellites and launch supply chains have drastically changed, and the price points have changed along with them. In 1991, there were less than 15 satellite manufacturers and a handful of launch manufacturers. By 2019, worldwide, there were over 2500 organizations that had either built and flown satellites or were involved in the business of building satellites. The supply chain for satellite components has gone from single-digit suppliers in the 1990s, where individual boxes, such as avionics, cost tens of millions of dollars and required years of lead time to today, where you can order an avionics board and add the software as an option from a website using a credit card to have the parts shipped to your home. The average price of a small satellite has dropped from $25M to less than $1M in two decades. The net result of this is a constellation of 100 satellites, which Iridium fielded, which has reduced implementation costs from $10B spent on the first Iridium constellation to less than $200M for a modern Internet Of Things constellation. This is much easier to finance at a lower CAPEX amount than the multi-billion-dollar constellations, and more market risk can be taken with a given constellation. The net result will be more innovation in the technology and the business cases. The market is placing enormous focus on leveraging the potential of nano and microsatellites to displace and disaggregate many of the missions currently taken up by larger satellites.

The current demand for launch capabilities vastly outstrips supply, and small satellite providers have shown a significant unmet market opportunity impeded by the need for more sufficient and flexible launch options. Around 1,500 small satellites are annually planned for launch in 2022-2023, according to a leading New Space market research group. Many constellation operators seek launch opportunities regularly, yet they often wait at least 9-18 months for a launch slot. According to Goldman Sachs, over 75% of existing new space companies say that their ability to generate revenue in the market is impeded by access to launch. Also, depending on the architecture, a constellation may demand a dedicated launch solution (not piggybacking via rideshare with a larger satellite) to optimally deploy into the correct orbit and at the right time to constitute the constellation. Euroconsult Research recently published a report, "Prospects for the Small Satellite Market," where "on average 580 small satellites will be launched every year (not counting SpaceX Starlink) by 2022 and 850 per year in subsequent years up to 2027". Broadband communications will be the main applications for the smallsats, with around 3,500 units from 2018 to 2027. Earth observation satellites will almost triple compared to the previous decade from 540 to 1,400 during 2018 – 2027". A dedicated launch is the most desirous launch strategy, albeit the most expensive, until smaller mass-manufactured launch vehicles come online.

The U.S. Government strongly supports the proliferation of the commercial launch sector. Space represents the ultimate high ground for the U.S. military and national security interests. Much of the military's assets are space-based in the form of satellites which are increasingly viewed as vulnerable to attack or damage. The unclassified and classified space portions of the U.S. DoD budget are expected to proliferate in the medium and long term. Although the DoD Space budget is broadly classified, Goldman Sachs estimates $22B in addressable annual spending, growing at a 6% CAGR. (Source: Goldman Sachs A&D Profiles in Innovation-Space 2016). Previous generations of satellites were developed in an environment where the U.S. assumed there would not be a reason to attack them. The resulting architecture has a small number of costly satellites with many capabilities packed onto each one. In response to this threat, the Office of Operationally Responsive Space was created as a joint initiative of several agencies within the U.S. Department of Defense to plan and prepare for the rapid development of highly responsive space capabilities (including launch) that enable the delivery of timely warfighting capabilities, and when directed by Joint Force commanders for on-demand space support, augmentation, and reconstitution. Through a range of programs, such as the DARPA

Launch Challenge, meant to foster responsive and flexible launch, the U.S. government seeks to establish a rapid-launch capability whereby it can replace satellite losses due to hostile action and reconstitute national capabilities using small satellites. With a $30B annual budget, NASA is a critical player in space and supports many private companies that have been shown to innovate market areas that NASA deems essential. Given the strategic importance of the space industry, the U.S. government is also supporting the private sector with critical legislative activity.

Conclusion

When Elon called me out of the blue in 2001, his stated purpose was a modest mission to Mars. I spent countless hours with Elon traveling around the world and discussing his goals for this mission and, correspondingly, his life philosophy. At the time, while this seemed lucrative and exciting, it did not seem historic. None of us involved in the early days of what became SpaceX ever imagined the global impact it would have. I did, however, have deep insight into Elon's dreams for Mars, and this has become increasingly important as time has marched on and has guided my predictions about his next moves. My years spent with the man gave me an evident insight into his thinking, and I use this insight to predict his next moves in human exploration and technological developments.

Even back in 2001, Elon was thinking about Mars bases. While we were working on Mars Oasis, Elon had secretly hired a small group of people to 'Imagineer' Mars bases. He shared some of those results with me at a meeting we had in Palo Alto in late 2001. I was stunned by what I saw, and Elon asked me what I thought. My response was simple: "Elon, we are having enough trouble with telling people that we are going to make our rockets go to Mars, let alone saying that we will be building Mars bases when we get there." I continued: "This is much like talking about aliens. We just can't do it publicly". Elon nodded in understanding, but I suspect he kept up a secretive effort to colonize Mars.

*The first humans to step foot on Mars will come
by commercial systems, not government built spacecraft.
Image Credit SpaceX*

Elon's actions since then have confirmed his long-term goal of colonizing Mars. One could be forgiven for arguing that the entire purpose of SpaceX is to ultimately colonize Mars. And they would not be wrong. Elon has made many public statements to that effect, making this idea less than speculation at this point. My deep understanding of this ultimate SpaceX goal has guided much of my advice to others trying to interpret their technology development efforts and product development paths. It has also guided my interpretations of the emerging market needs in the launch and whether SpaceX would efficiently service those markets.

SpaceX made space safe for investing, and as a result, we are in the midst of an unprecedented expansion of the space economy. I, for one, believe that this new chapter in space exploration will inspire people to imagine a more transparent future where information dark ages are no longer possible under repressive regimes, where we as a species understand our planet better and can predict its behavior better and where humankind has begun to take its first feeble steps into the cosmos in permanence. This bold vision for the world was long in the minds of us Space Cowboys back when we emotionally and intellectually split from our government space efforts 30 years ago. It's only a matter of time before it now becomes a reality.

Despite the importance of Elon and countless other billionaires who have made undeniable imprints on the space industry, in the end, it comes down mainly to the efforts of individuals from all walks of life. Despite being led most visibly by billionaires such as Elon Musk and Jeff Bezos, this revolution has been accomplished, at its core, by ordinary people with extraordinary drives who have somehow come together through serendipity and hard work to create a bright future without the lead of nation-states and without waiting for others to make these dreams a reality. This New Space Race belongs to us all. Its future is in our hands. Make of it what you can and own the future.

ABOUT THE AUTHOR

Jim Cantrell is an author, inventor, entrepreneur, and founder of numerous space startups. He is considered a luminary in the aerospace industry with more than 30 years of proven leadership experience and significant contributions to space technology. Jim also served as a founding team member at SpaceX.

Printed in the USA
CPSIA information can be obtained
at www.ICGtesting.com
LVHW041458191023
761550LV00054B/721